ARTIFICIAL REALITY

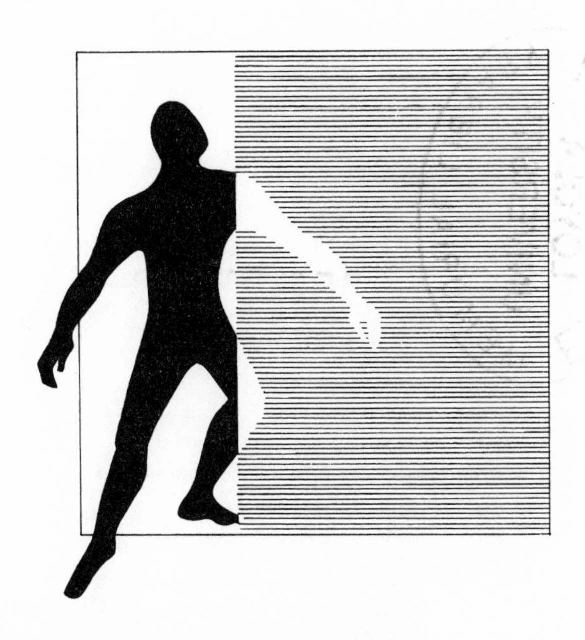

ARTIFICIAL REALITY

Myron W. Krueger

University of Connecticut

 *ADDISON-WESLEY PUBLISHING COMPANY*
Reading, Massachusetts Menlo Park, California
London Amsterdam Don Mills, Ontario Sydney

The donations of the following companies to VIDEOPLACE are greatly appreciated.

American Microsystems, Inc., Santa Clara, CA
AMP, Inc., Harrisburg, PA
Augat, Inc., Attleboro, MA
Diablo Systems, Inc., Hayward, CA
Gardner Denver Co. Pneutronics, Grand Haven, MI
Hewlett-Packard, Corporate Div., Palo Alto, CA
Hughes Aircraft Co., Los Angeles, CA
Lambda Electronics, Corpus Christi, TX
3M Co., St. Paul, MN
Micro Networks Corp., Worcester, MA
North Electric Co., Galion, OH
T&B/Ansley Corp., Los Angeles, CA
TRW Cinch Connectors, Elk Grove Village, IL
Woven Electronics, Mauldin, SC

Library of Congress Cataloging in Publication Data

Krueger, Myron W.
Artificial reality.
1. Technology and the arts. 2. Man-machine systems.
3. Avant-garde (Aesthetics) I. Title.
NX180.T4K7 700'.1'05 82-3897
ISBN 0-201-04765-9 AACR2

ISBN 0-201-04765-9
ABCDEFGHIJ-DO-898765432

To Joan, Mike, and Kristi

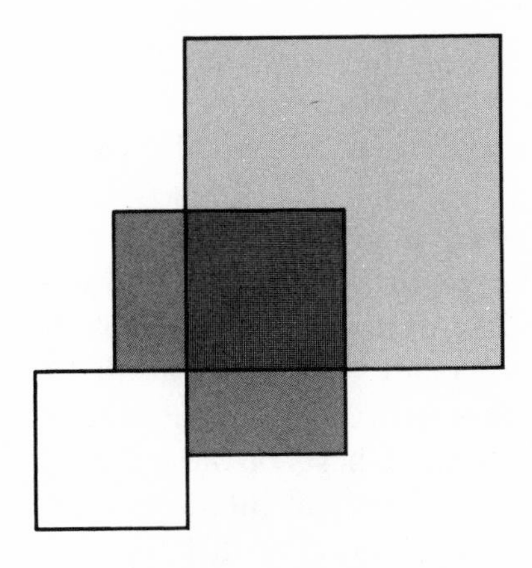

PREFACE

The central idea of this book is the Responsive Environment, a paradigm like Hobbes' model of the world as a huge machine, which aids in understanding the essence of our current and future experience. This experience will be characterized by computer systems able to sense our needs and respond to them.

Unlike other books that deal with the implications of computers, *Artificial Reality* is not a survey of existing applications. Rather, it describes GLOWFLOW, METAPLAY, PSYCHIC SPACE, and VIDEOPLACE, Responsive Environments that provide glimpses of what the future may be like. Its intended reader is anyone with an interest in art or technology who is open to some unexpected ideas about how the two may intersect. This reader might be an artist or a technologist or might have an interest in video, psychology, or education. In fact, the reader could be anyone who likes to keep up with new ideas.

The book is roughly divided into three sections. The first section, including Chapters 1–4, describes interactive environments that perceive human activity and respond through intelligent visual and auditory displays.

The second section, Chapters 5–8, gives an aesthetic and technical definition of Responsive Environments. The aesthetic discussion evolves from the premise that a technology that pervades our lives must be dealt with as an aesthetic issue as well as an engineering problem. Chapter 9 introduces VIDEOPLACE, which represents a synthesis of these ideas.

Chapters 10–12 forecast applications of interactive environments and reflect on the implications of artificial reality.

While the book does contain some discussion of technical issues, every effort has been made to present this information in an intuitive way. For the technical reader, or the non-technical reader with an interest in learning more, tutorial information and additional technical details are provided in the appendices and glossary.

ACKNOWLEDGMENTS

Artificial Reality is the product of many hands. The long-term editorial assistance of Joan Sonnanburg was essential to writing the book. She also created the diagrams.

Over the years, several hundred people have contributed directly or indirectly to the work described in *Artificial Reality*. Some individuals volunteered thousands of hours of their time, while others provided institutional support. The original encouragement and support for the book came from my doctoral advisor, Leonard Uhr, at the University of Wisconsin and from a Research Assistantship under an NSF grant.

Later, Tom Haig and Verner Suomi provided an institutional umbrella at the Space Science and Engineering Center of the University of Wisconsin. Gilbert Hemsley provided assistance in securing financial support. The National Endowment for the Arts, the Brittingham Trust, and the Wisconsin Arts Board provided funding during this period.

For the last three years, support has come from the Electrical Engineering and Computer Science Department at the

University of Connecticut and from the Connecticut Research Foundation. There have also been donations of equipment and components by corporations. (See copyright page.)

Of the many individuals who have participated, several deserve special mention. Keith Sonnanburg provided valuable comments on the manuscript's structure. In METAPLAY and PSYCHIC SPACE, Wayne Weber was responsible for the electronics and computer interfacing. At the Space Science Center, Chas Moore was instrumental in pursuing arts funding, while Joan Sonnanburg obtained industrial support. John Beetem assisted with the technical work during this period. Finally, at the University of Connecticut, Victor and Peter Odryna have made significant contributions to the operation of the VIDEOPLACE laboratory.

Storrs, Connecticut M. W. K.
1981

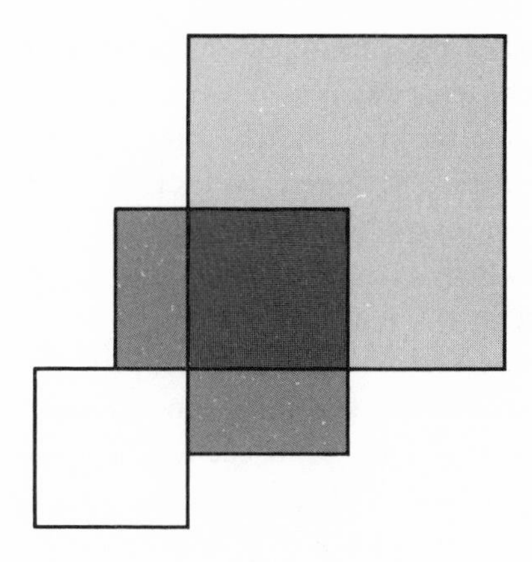

INTRODUCTION

The world described in *Genesis,* created by mysterious cosmic forces, was a volatile and dangerous place. It molded human life through incomprehensible caprice. Natural beneficence tempered by natural disaster defined reality. For centuries the goal of human effort was to tame Nature's terrible power. Our success has been so complete, that a new world has emerged. Created by human ingenuity, it is an artificial reality.

Our daily experiences are overwhelmingly formed by this reality. It is the automobile, the written word, television, and the temperature-modulated building, not the natural environment, that influence our lives. Each day we come closer to controlling every aspect of the human environment. Ironically, we are profoundly ignorant of what we have made.

This ignorance spawns an understandable malaise. However, it is as foolhardy to yearn for a benign Nature, which never existed, as it is to accept technological developments that make us uneasy. Rather, we must explore ways of shaping our environment that lead to psychological comfort and aesthetic pleasure as well as efficiency.

The practical concerns of technology and the humanistic concerns of aesthetics need not be estranged. In fact, during most of human history, technology made life more humane. Artists were conversant with the advanced technology of their culture and employed it in their art. However, the unprecedented rate of scientific and technological developments during the past century created a chasm between science, technology, and aesthetics. Today, however, these disciplines are coming together again with considerable force. This trend is part of a larger cultural implosion that is just beginning — the integration of all aspects of society by interconnected information, communication, and control systems. These networks and the computational power they bring will permeate our lives much as electricity does today. Cybernetic systems will sense our needs and enter our offices, homes, and cars. We will live in *Responsive Environments.*

The focus of this book is the interaction of people with machines, in both the immediate interface and in the broader cultural relationship. In spite of the fact that a significant fraction of the work force spends an ever increasing percentage of its time dealing with machines, particularly computers, we rarely ask: what are the ways in which people and machines might interact and which of these are the most pleasing? Here, the question is posed as a problem in aesthetics. The tool for exploring this issue is the computer-controlled Responsive Environment, which serves as a paradigm for our future interaction with machines.

In this book a Responsive Environment is an environment where human behavior is perceived by a computer, which interprets what it observes and responds through intelligent visual and auditory displays. Since many kinds of displays, including discrete lights, video, computer graphics, and electronic music are amenable to computer control, a rich repertoire of relationships can be established between an individual and the Environment. The Environment can be completely controlled by a preexisting program, or operators can intercede and use the computer to amplify their ability to interact with other people. In either case, a programmer anticipates the participant's possible reactions and composes different response relationships for each alternative. The program is a composition that can respond and learn on its own after it is completed.

The Responsive Environment was conceived as a new art form. It represents a unique melding of aesthetics and technology in which creation is dependent on a collaboration between the artist, the computer, and the participant. The artist composes a network of intelligent response relationships. The participant explores this universe, initially triggering responses inadvertently, then gradually becoming more and more aware of causal relationships. The computer perceives and interprets the participant's actions and responds intelligently. The art form is the composed interaction between human and machine, mediated by the artist.

While conceived with aesthetic intent, the implications of the Responsive Environment go beyond art. The Responsive Environment is a generalized facility that separates technology from any single application, enabling us to examine its broad implications. Thus, we can judge it in isolation from purely practical concerns or use it to expand our understanding of specific fields.

The first four chapters of this book trace antecedents of the Responsive Environment, discuss related trends in art and technology, and describe three early manifestations of the Responsive Environment: GLOWFLOW, METAPLAY, and PSYCHIC SPACE. Chapter 5 discusses the role of the Responsive Environment as a new aesthetic medium. Chapters 6 through 8 explain how Responsive Environments work and discuss what issues should be addressed in their design. Chapter 9 introduces VIDEOPLACE, an extremely versatile and generalized Responsive Environment which is currently under development. The closing chapters forecast how these analogues of artificial reality will be used in computer science, education, psychology, the arts, and in our daily lives.

This book is the outgrowth of thirteen years of research dedicated to affirming the beliefs that humanity's relationship to technology can be a positive experience and that in a world in which technological innovation increasingly defines our experience, the quality of our interactions with machines will largely define the quality of our lives. By making technology both palpable and palatable, the Responsive Environment seeks to engage people in a playful exploration of the coming fact.

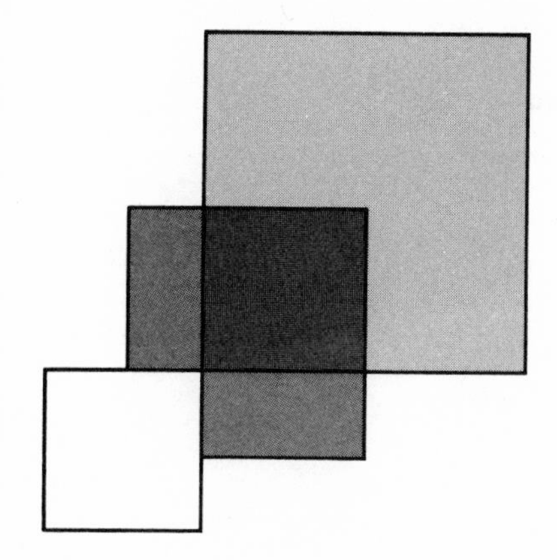

CONTENTS

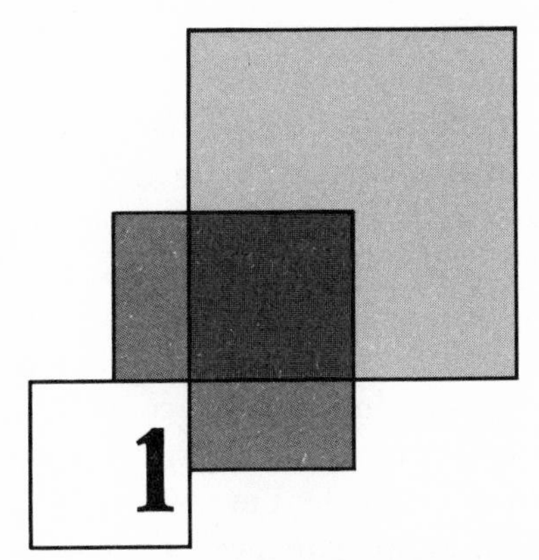

ANTECEDENT TRENDS IN ART AND TECHNOLOGY

ANTECEDENTS

Artists today are not to be confused with technologists. Most would be appalled if they were. Yet, this was not always the case.

Art and technology used to be inextricably intertwined. Prehistoric artists were using state-of-the-art technology when they ground their pigments. From the Middle Ages through the Renaissance, art was the most powerful means of communicating the beliefs of a culture to a largely illiterate public. The architects who built awesome cathedrals and the painters who invented systems for realistically depicting perspective were well versed in the technology of their day.

As the predominant Western world view slowly shifted from a belief in religious truths to a belief in scientific explanations, developments in science and technology heavily influenced artists. In the nineteenth century, the development of the camera freed the painter from the role of visual historian, and research on the nature of perception prompted

Impressionist painters to make pictorial explorations of how and what we see.

In the twientieth century something unusual happened. Scientific discoveries and technological developments proliferated at an unprecedented rate. The scientist became absorbed with the ultrafast, the ultrasmall, the ultraslow, the infrared and the ultraviolet, all invisible, to be dealt with as conceptual abstractions that could be experienced only by inference and calculation. Into this realm the artist could not follow. Without becoming a scientist or technologist, the artist could no longer keep abreast of all the new events. Although perhaps related at a philosophical level, the communities of the artist and the scientist became increasingly alienated. Both agreed that there was no science in art and no art in science.

During the past two decades, there have been developments that herald the return of a relationship among art, science, and contemporary technology; for example, the efforts of artists to utilize today's most potent means of communication: video and the computer; of scientists to represent abstract concepts visually; and of technologists to increase the possibilities of telecommunication and simulated experience. The purpose of this chapter is to consider these conceptual antecedents of the Responsive Environment in science and art.

DEVELOPMENTS IN SCIENCE

There is a considerable amount of effort in the scientific world directed at developing richer interactive systems. Whether the intent is communication or simulation, there is a common goal of creating ever more convincing representations of reality or of hypothetical structures. This scientific interest presages an interest on the part of our entire culture. The pseudoexperience systems that are evolving will be applied to education, psychology, art, and telecommunication.

Communication or Coincidence

An interesting aspect of communication that is seldom noted can be illustrated as follows. A number of years ago, whenever one of my son's friends called him on the phone, he would ask "Is Mikey here?" Being in command of the facts of geography, adults usually think of communication as the transmission of information from one point to another. Children, on the other hand, believe that if they can talk to someone, they must be in the same place. In other words, our concept of "place" is based upon the ability to communicate. The place created by the act of communication is not necessarily the same as that at either end of the communication link, for there is information at each end that is not transmitted. The "place" is defined by the information that is commonly available to both people.

There is a definite trend towards expanding the sense of being in the same place. We can see this in the development of transmission systems from Morse Code to the telephone, to radio, to black-and-white television, and finally to color television. Each of these broadcast and dissemination systems allows us to perceive events from afar more completely than its predecessor. Systems that are now being developed will allow us not only to perceive distant events, but also to act at a distance. For example, ground-based fighter pilots can already fly airplanes that are faster and more maneuverable than they would be if a pilot were aboard; and Russian scientists while sitting on earth can explore the moon via their remotely controlled lunar vehicle.

Super Senses

We are no longer creatures of five senses. Technology has given us hundreds. We can sense the universe throughout the electromagnetic spectrum. We can hear vibrations, from the infrasound of the seismologist to the ultrasonics used in destructive testing. We can see molecular and cosmological structures. We can sniff the stars through spectral analysis. We can feel the age of ancient objects. But, in every case, we must convert the data from these new senses into a form that our original five senses can understand.

The field of medicine has made some startling contributions in this area. Physicians have always longed to see inside the body without doing injury to it. A number of such noninvasive techniques have been developed; in each case, however, there has been the need to reconstruct this complex data in a form that can be grasped intuitively by the doctor. The thermographic camera, used as a diagnostic device, produces a color video image of the heat radiating from a patient's body by assigning different colors to different temperature ranges. Since different types of tissue have different temperatures, this camera is useful for locating tumors, which tend to be warmer than surrounding tissue.[1]

Ultrasound imaging, initially devised for radar and sonar, is now used to explore within the human body without ill effects. In obstetrics, ultrasound has virtually replaced other methods; it is also widely used in cardiology.[2] Computerized tomography is yet another means of representing previously hidden information. This technique makes visible a two-dimensional slice of the body, enabling a doctor to make subtle discriminations such as distinguishing types of brain tissue from each other or distinguishing normal blood from coagulated blood.[3] With such techniques, it will ultimately become routine for a doctor to look at live displays of a fetus or a beating heart. The doctor will be able to explore the inside of the body visually as easily as the outside.

Conception Becomes Perception

Lately, there has been a renewed appreciation of our intuitive intellect, the one with which we apprehend the everyday world of physical objects. Scientists recognize that they can better understand concepts if they are visible and can be manipulated. In other words, concepts that can be physically experienced are more easily understood than those which can only be thought about. To this end, chemists and technologists have developed techniques for displaying complex chemical models in three dimensions using interactive computer displays that permit scientists to manipulate the computer model just as they used to twist and turn the old physical models made from sticks and balls.[4] Computer animation techniques are also giving astronomers the chance to study the dynamics of galaxies over billions of years.[5]

Understanding the value of complete interactive representations of reality has led to the development of 3D graphic systems which allow the user to visually explore a three-dimensional design space.[6] Mathematical functions are commonly shown as three-dimensional surfaces. Even abstract mathematicians, such as topologists, find that visualization can spur conceptual thinking. Since we evolved to cope with three-dimensional spaces, we are incapable of visualizing the higher dimensional systems that our mathematicians create. To combat this problem, Michael Noll at Bell Labs has been using stereo computer graphic images to try to experience four-dimensional systems by exploring their projections onto three-dimensional space.[7]

Ultimately, our representations of the unseeable and the conceptual will evolve to a point where we can not only see them, but hear, feel, and walk around in them as well.

Human-Machine Systems

Human-Machine interaction has progressed a long way since the 1950s when O. K. Moore used humans to simulate machines in early versions of responsive environments. These were designed to teach young children to learn.[8] Today, the human and the machine are joined to form a single functioning unit, a symbiotic entity in which both perform the tasks suited to their abilities. Such interaction is not only the dominant mode for using computers, it may become the dominant mode for working in general.

Simulation Is Artificial Experience

As the tools of science become more expensive and the tasks to which they are set more ambitious, there is increasing pressure to minimize the risks of operator error and to anticipate effects of bad design by rehearsing the situation as fully as possible through simulation.

Electronic circuits, mechanical systems, chemical factories, and military strategies are all tested first in the computer. Medical and business students can see the effect of their decisions on simulated cases. Flight simulators that are used to train both military and commercial pilots give complete representations not only of relevant information, but of the whole experience.[9]

These systems generate dynamic, three-dimensional, colored, and textured images in real time. NASA used similar systems to simulate moon landings and to investigate human-factors issues involved in loading and unloading the space shuttle in orbit, so problems could be corrected before the real shuttle was launched.[10] One of the most elaborate simulations to date is one developed by General Electric to train oil tanker captains. It uses seventeen color video projectors to simulate what a captain would see in every direction from the bridge of the ship as it enters New York's harbor.[11]

In order to function, these simulations must sense our actions and respond appropriately. This aspect of our synthetic realities will be incorporated into our everyday environment, making it increasingly artificial, almost a simulation of what we would like the world to be.

The Environment Senses Our Needs

Systems that can sense our actions will also be able to infer our needs and respond to them. Simple examples of such environmental systems that the layperson encounters today are the thermostat in the home and the automatic door at the grocery store. The minimal requirement for this type of system is a sensor that determines when its function should be performed. We not only can expect to find automatic sensing systems in more of the technology we encounter, we already have traffic sensors that merge one line of traffic with another, ultrasonic burglar alarms, and automated battlefields. It is easy to anticipate devices that sense our presence in a room and regulate the sound, heat, and lighting accordingly. We will live in Responsive Environments.

As these isolated devices become more sophisticated and are integrated to form systems with the ability to move and speak, we will be faced with the presence of the computer in our lives as an apparently living force. What we are now witnessing is a birth process — the birth of the artificial entity as an integrated, perceiving, behaving system. Whether or not these entities will evince intelligence is a separate issue; the artificial entity is inevitable. The early forms are already here.

EXPANDING ART

The development of new technologies has been paralleled by the invention of new aesthetic concepts. Beginning with the Dada movement in the early twentieth century, the traditional assumptions about what art is have been continuously challenged. New ideas and new forms have been tried and discarded or filed away for future reference. However, artists have been more interested in exploring immediate possibilities than in laying the groundwork for a new medium. Their lack of technological skills and, more significantly, their lack of empathy with technology itself have often prevented them from seeing that their early insights contained the basis for a new aesthetic tradition.

Surrendering Control

As artists have discarded traditional forms, new attitudes have discreetly begun to appear. Most important is a feeling among artists that they are no longer necessarily the creators of objects. They are no longer to be judged exclusively by their command of a medium. In fact, in many cases, it is the artist's willingness to forego control that constitutes a contribution. The theme of surrendering control to influences within or beyond the artist recurs in recent theorizing motivated by a desire both to discover new kinds of order and to involve an audience.

The composer John Cage has been a major influence on this school of thought. He advocated a search for new sound patterns based upon the translation of all sorts of relationships drawn from other areas. While others use the word "randomness" when describing his method, Cage himself speaks of "unintended" sound.[12] He seems to say that he is bored with the sounds that he can make intentionally. If he can conceive the sound in his mind, knows how to realize it, and will not be surprised when he hears it, he has learned nothing. The whole exercise would be a waste of time. Therefore, he suggests that a musician become a more sensitive listener, surrendering to forms of order beyond those already explored. To find new forms, one must give up the kind of control that was learned from the old ones.

Happenings

Cage's ideas of randomness and spontaneity influenced the Happenings of Allan Kaprow in the 1960s.[13] A Happening was theatre without an audience. Nothing was conceived with the passive spectator in mind. A very loose series of possibilities were planned; the participants were the ones who actually give the work its final form. Here, as in Cage's work, the artist surrendered immediate control, stepped back to a higher level, and gave the actors and the audience a level of control heretofore unknown. Responsive Environments also require the artist to accept reduced control, to think in terms of a structure of possibilities that leaves the final realization of the piece in the hands of each participant.

Conceptual Art

We usually think of an artwork as a physical object that we perceive directly through our senses. In conceptual art there may be objects or events that are the physical manifestations of a piece, but that have little aesthetic significance in themselves. The art lies in the concepts that are suggested by these outward signs and explained by written or graphic documentation. The perception that is evoked is a product not of our senses, but of our intellect. Thus, conceptual art emphasizes the appreciation of ideas as a source of aesthetic experience.[14]

Process Versus Object Art

Perhaps one reason that process as art has gained favor is that craftsmanship has been devalued by machines that can produce slick objects in a fraction of the time it would take an artist. When objects are part of today's artistic process at all, they are often intended only as clues to an intelligent process of aesthetic exploration.

This attitude is appropriate. The artist's objective today should be experimentation with new modes of thought and new technology. When a degree of mastery is achieved, the results can be tested against whatever criteria still seem relevant.

Art and Technology

In recent years, artists and technologists have embraced the new tools and sought to create a new type of imagery. Computers, video, film, and electronic music have been wedded in a variety of experimental forms.

Video art has broken with television's traditional role as a broadcast medium and exploited its techniques to create a new art form. Video technology provides a variety of new image-processing techniques. It also allows the recall of images, both for instant replay and for long-term storage. While commercial television usually uses video as if it were film, video art attempts to exploit the medium's unique characteristics.

The early work in computer art was based upon batch processing rather than an interactive model of programming. In all cases, the task was specified and programmed, and the output was the composition. More recent systems allow an artist to explore a line of development, intervene at every step, judge the outcome, and then use it or discard it. The computer should increase the artist's ability to experiment so that the final compositions are based upon a rich experience with the medium and the full expression of the artist's own ideas.

Computer-Controlled Responsive Art

Much of the art that has been produced with the computer could have been produced by other means, albeit tediously. True computer art would be impossible without the computer. One essential aspect of the computer is that it can assimilate information and make decisions in real time. In the past decade, there have been some attempts to create an art form dependent on this computer capability. While there are only a few such works, they point to a major new thrust in both technology and art. For example, since 1970 the author has explored the interactive capabilities of both video and computer technology.

These steps towards a collaboration between art and technology represent a positive trend. An artist who is alienated from technology cannot speak for a technological culture, any more than a technologist who disdains aesthetics can design a humane technology. The Responsive Environment has its roots in this confluence of science, technology, and art.

The ideas presented in this chapter are the conceptual framework for the author's work. The rest of the book describes an intellectual quest with a number of simultaneous goals: first, the aesthetic and technical development of a powerful new artistic medium; second, the full expression of this medium for practical as well as aesthetic purposes; and last, the definition of a humanism that recognizes technology as an important part of human nature, not as an external force.

NOTES

1. Thermovision, AGA Corp., Secaucus, New Jersey.

2. G. B. Devey & P. N. T. Wells, "Ultrasound in Medical Diagnosis," *Scientific American* 238 (May 1978):98–104.

3. W. Swindell & H. H. Barrett, "Computerized Tomography: Taking Sectional X Rays," *Physics Today* 30 (December 1977):32–41.

4. J. S. Lipscomb, "Three-Dimensional Cues for a Molecular Computer Graphics System," doctoral dissertation, University of North Carolina at Chapel Hill, 1979.

5. A. L. Robinson, "Computer Films: Adding an Extra Dimension to Research," *Science* 200 (May 1979):749–52.

6. Frederick P. Brooks, "The Computer Scientist as Toolsmith — Studies in Interactive Computer Graphics," (International Federation of Information Processing Societies, 1977).

7. A. Michael Noll, "Computer Animation and the Fourth Dimension," Fall Joint Computer Conference (American Federation of Information Processing Societies), 33–2 (1968):1279–83.

8. O. K. Moore, "O. K.'s Children," *Time* 76 (November 7, 1960):103.

9. Barry Miller, "Remotely Piloted Aircraft Studied," *Aviation Week and Space Technology* 92, No. 22, (June 1, 1970):14–15.

10. R. Weinberg, "Computer Graphics in Support of Space Shuttle Simulation," *ACM-SIGGRAPH Proceedings* 12, No. 3 (1978):82–86.

11. "Getting There," *Electronics*, September 30, 1976.

12. John Cage, *Silence* (MIT Press, 1967), pp. 8–11.

13. Henry Geldzahler, "Happenings: Theater by Painters," *Hudson Review* 18, No. 4 (Winter 1965–66):65–66.

14. Lucy Lippard, *Six Years: The Dematerialization of the Art Object. from 1966–1972* (Praeger, 1973).

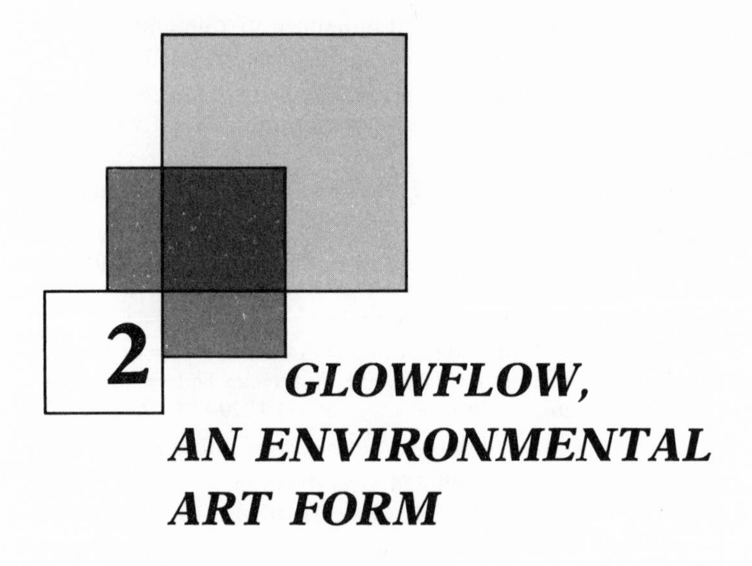

2 GLOWFLOW, AN ENVIRONMENTAL ART FORM

BACKGROUND, CONCEPTION, AND DESCRIPTION

In early 1969, I became involved in the development of GLOWFLOW, a computer art project envisioned by Dan Sandin, a physicist who has since become an artist; Jerry Erdman, a minimalist sculptor; and Richard Venezsky, a computer scientist. GLOWFLOW was conceived as an encounter between art and technology and was exhibited to the public at the Memorial Union Gallery of the University of Wisconsin in April 1969.

GLOWFLOW was an aesthetic light-sound environment, controlled by computer, with a limited provision for responding to the people within it. While the original plan was a group effort, the visual and musical designs developed separately. Jerry Erdman designed a darkened room in which a visual display created a perceived space only partially faithful to the actual physical space. Changes in the display were indirectly contingent upon the actions of participants. This part of the design was deliberate; it was felt that direct responses would become a focal point reinforcing noise in what was intended to be a quiet, contemplative environment.

The physical space was an empty rectangular room constructed within the gallery (Fig. 2.1). The display consisted of a suspension of phosphorescent particles in water, pumped through four transparent tubes attached to the gallery walls. Each tube contained a different colored pigment. Since the room was dark, the lighted tubes provided the only visual reference and were arranged to distort one's perception of the room, causing the room to appear wider in the center than at the ends (Fig. 2.2). The disorienting darkness also caused the viewer to assume that the bottom tube was level with the floor. Thus, when people discovered the floor line moving higher with respect to their own position, they thought they were going downhill. This illusion was so strong that people actually leaned backwards as they moved.

The tubes were run through opaque columns along the walls, each of which contained four lights, one for each tube passing through it (Fig. 2.3). When a light was turned on, the phosphorescent substance flowing through the corresponding tube would glow as it emerged from the column. Its intensity decreased as the phosphorescent fluid moved along the wall until the light finally decayed approximately twenty feet from the orginating column.

Fig. 2.1 Gallery floorplan for GLOWFLOW

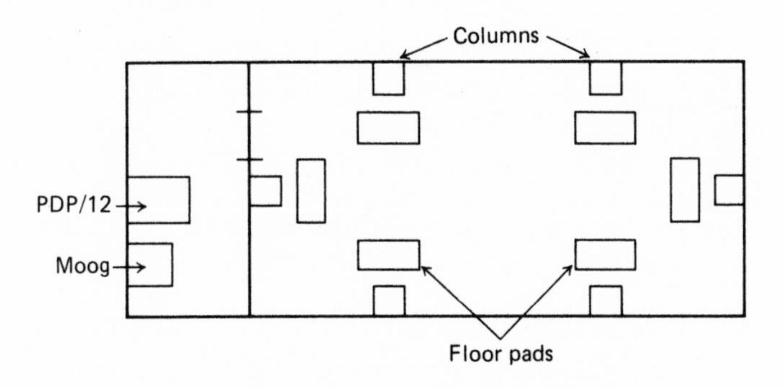

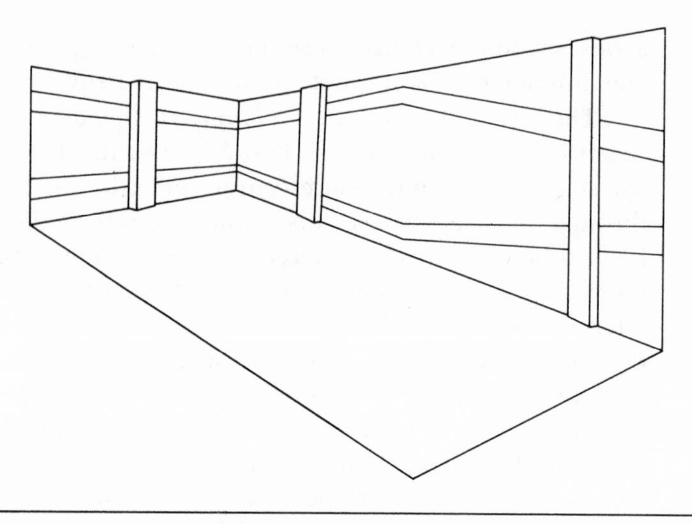

Fig. 2.2 GLOWFLOW tubes on gallery wall

There were six columns, two along each of the side walls and one on each end wall. With four lights per column, there were a total of twenty-four lights that could be turned on or off in any combination by the computer. Thus, it was possible to light all of the tubes coming out of a single column, a single tube as it came out of every column, or a random combination of tubes and columns.

Fig. 2.3 System for activating phosphors

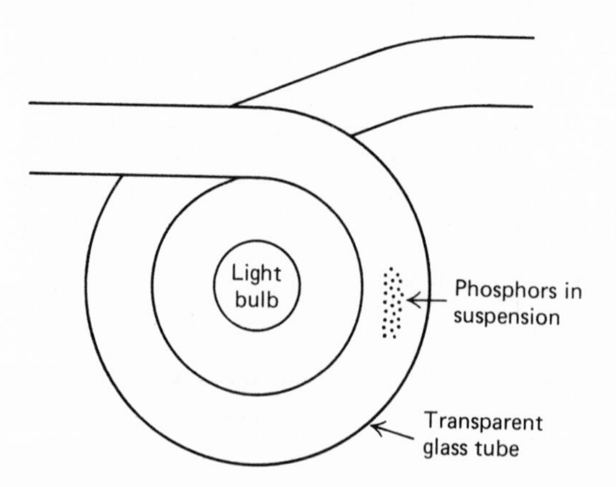

ELECTRONICS

The environment was controlled by a PDP-12 minicomputer manufactured by Digital Equipment Corporation. The interface consisted of three sections: one for speakers, another for a Moog synthesizer, and a third for lights.

The PDP-12 controlled the Moog synthesizer via the interface. The Moog sounds were fed through amplifiers to the speaker in each of the columns. It was possible to feed sound through any combination of the six speakers. Sounds could rotate around the room or bounce from one wall to the other.

The only sensing devices were pressure-sensitive pads in front of each of the six columns. In each of these pads was a switch that could be sensed by the computer. The pads were used to sense the presence of a person standing on them.

THE GLOWFLOW EXPERIENCE

Between fifteen and twenty people were in the darkened room at any one time. New people were allowed in as others left. There tended to be several stages in each person's experience of the environment. First, there was a disorientation due to the darkness. During this period people stayed near the entrance. As their eyes grew accustomed to the low light level they would explore the room, discovering the illusory nature of the perceived space. Later they might sit or lie down on the floor, interacting with other people or in quiet contemplation. As long as there was some turnover of people, no one was pressured to leave the room.

People had rather amazing reactions to the environment. Communities would form among strangers. Games, clapping, and chanting would arise spontaneously. The room seemed to have moods, sometimes being deathly silent, sometimes raucous and boisterous. Individuals would invent roles for themselves. One woman stood by the entrance and kissed each man coming in while he was still disoriented by the darkness. Others would act as guides explaining what phosphors were and what the computer was doing. In many ways the people in the room seemed primitive, exploring an environment they did not understand, trying to fit it into what they knew and expected. Since the GLOWFLOW publicity mentioned this responsiveness, many people were prepared to experience it and would leave convinced that the room had responded to them in ways that it simply had not. The birth of such superstitions was continually observed in a sophisticated university public.

INTERACTIVE DILEMMA

The artists' attitude toward interaction between the environment and the participants was ambivalent. Responsive relationships were seen as conceptually interesting but the artists did not feel that it was important for the audience to be aware of them. The idea of direct response to movement and voices was discarded. It was feared that if immediate responses were provided, the participants would become excited and think only of eliciting more responses. This active involvement would conflict with the quieter mood established by the softly glowing walls. The power of responsiveness was recognized but avoided as not in keeping with the predominantly visual conception.

The environment responded to people in various ways. For instance, a sound might rotate around the room if a person stood or sat on a certain pad, or the pattern of lights might change. However, people had little sensation of interaction, for several reasons. First, there were many programmed delays between action and response. This time lapse prevented any awareness of the causal relationships. Second, the large number of people in the room meant that any response could have been elicited by someone else's action. Finally, the medium of glowing and flowing was itself slow to respond; it required seconds for a glowing line to appear and decay. Thus, if a person did cause a response, it was impossible to repeat it immediately to verify causality. With such intermittent responsiveness, it was impossible to establish an intimate relation between action and display. While GLOWFLOW was quite successful visually, it was precisely its emphasis on the visual conception that limited the impact of responsiveness.

The music in GLOWFLOW was never heard in its entirety due to hardware difficulties. I implemented the program planned by Madison composer Dr. Bert Levy, which used a stochastic model for the generation of the music and added provisions for a very rudimentary responsive capability. A more detailed description of the music is given in Appendix I. The program, while primitive, presented an interesting first step for responsive sound, a new discipline in itself.

LESSONS FROM GLOWFLOW

GLOWFLOW succeeded as a kinetic environmental sculpture rather than as a Responsive Environment. However, the GLOWFLOW experience led me to a number of conclusions:

1. Interactive art is potentially a very rich medium which must be judged on its own terms.

2. In order to respond intelligently, the computer should perceive as much as possible about the participants' behavior.

3. In order to highlight the relationships between the environment and the participants rather than among participants, only a small number of people should be involved at a time.

4. Participants should be aware of how the environment is responding to them.

5. The choice of sound and visual response systems should be dictated by their ability to convey a wide variety of conceptual relationships. The tubes of GLOWFLOW did not have a sufficient variety of responses. They represented a single visual statement rather than providing a medium of expression.

6. The visual responses should not be judged as art, nor the sounds as music. The only aesthetic concern should be the quality of the interaction. The interactive experience may be judged by very general aesthetic criteria: the ability to interest, involve, and move people; to alter perception; and offer a unique kind of beauty.

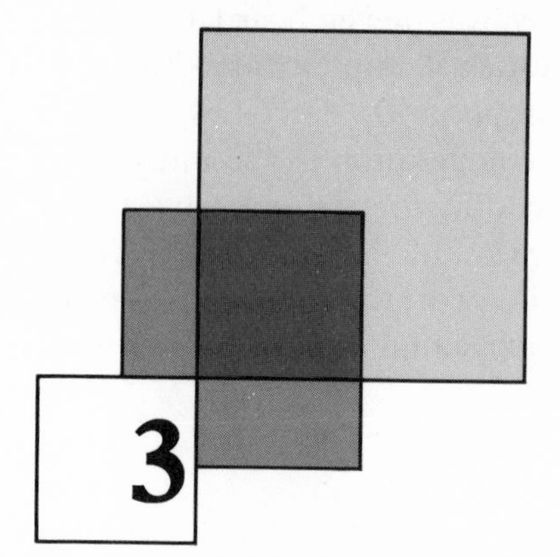

METAPLAY, COMPUTER-FACILITATED INTERACTION

BACKGROUND, CONCEPTION, AND DESCRIPTION

Following the GLOWFLOW experience, I conceived and directed METAPLAY which was exhibited at the Memorial Union Gallery in May 1970. The concerns of METAPLAY were a radical departure from GLOWFLOW, in which the visual, sound, and responsive issues were approached separately. METAPLAY sought to integrate these elements in a single framework. Traditional criteria of art, beauty, and responsive subtlety were set aside. The focus was on the interaction itself and on the participant's awareness of the interaction. METAPLAY investigated aesthetic interactions, not merely simple responses, but the development of complex, sophisticated responsive relationships. A commitment was made to personalize the experiences for each participant. This goal required that a number of independent hardware-software systems be implemented. In addition, the METAPLAY Environment was designed with the realization that the complexity of the equipment made intermittent partial failures inevitable. This fact mandated the use of several interlocking themes, guaranteeing that at least part of the exhibit would always be functional.

The physical environment was an empty, squarish, darkened room with one wall dominated by an 8′ × 10′ projection screen. This translucent screen allowed the video projector to be placed outside the Environment and to cast its image from the rear. The benefits of rear projection were that the projector was not an obtrusive object in the space and the viewer could not block the projection beam (Fig. 3.1). The remaining walls were painted with phosphorescent pigment. A large sheet of black polyethylene on the floor concealed 800 pressure-sensitive switches.

Two approaches were taken to the interaction. In the first, the computer was used to create a unique real-time relationship between the participants and a human facilitator, in another building. The live video image of the participant and a computer graphic image drawn by the facilitator were superimposed and rear-projected on the screen at the end of the gallery space. The viewer and facilitator both responded to what they saw on the screen.

Fig. 3.1 METAPLAY

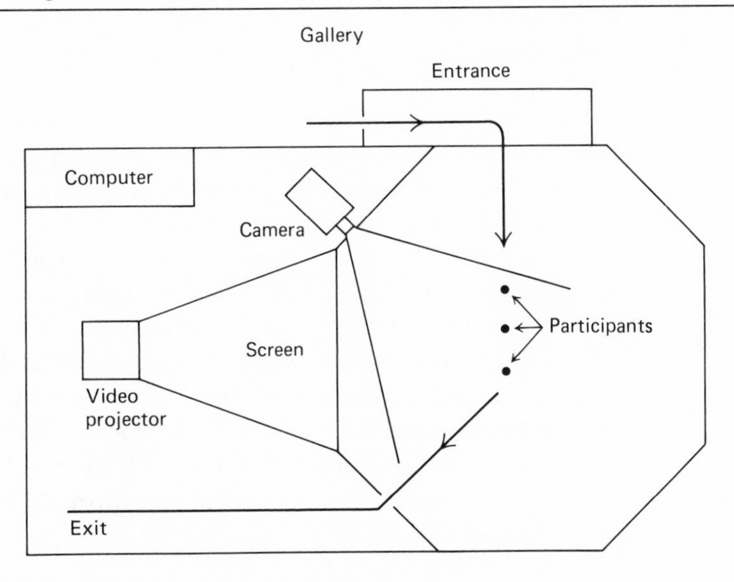

In the second approach a series of simple feedback relationships between the viewer and the computer were under automatic computer control. As the viewer moved around the Environment, the computer responded with electronic sound and projected graphic images.

A commanding real-time display was needed to express the interactive relationships. The most versatile existing real-time displays are the graphic display computer and closed circuit video. The problem with both of these media is one of scale; the standard 25″ monitor is too small to be an environmental display because we are conditioned to sit and watch rather than interact with it physically. The solution was to convert the computer image to video and rear project it on the 8′ × 10′ screen using a video projector. Live and computer-generated images could be shown individually or with the computer image superimposed on the live image. The result was more than an effective display: the marriage of computer and video images was a significant step in the development of the Responsive Environment.

COMPUTER-FACILITATED INTERACTION

Hardware

The image communications link began with a data tablet, an instrument that enabled the facilitator to draw or write on the computer screen (Fig. 3.2). The term facilitator is used instead of artist because the person at the data tablet did not always possess any recognized artistic skills. One video camera, in the university's Computer Center, was aimed at the display screen of a graphic display computer manufactured by Adage Corporation. A second camera, a mile away in the gallery, picked up the live image of people in the room. A television cable transmitted the video computer image from the Computer Center where it was displayed on a video monitor providing feedback for the facilitator who could speak to the cameraman in the gallery through the audio channel associated with the video signal.

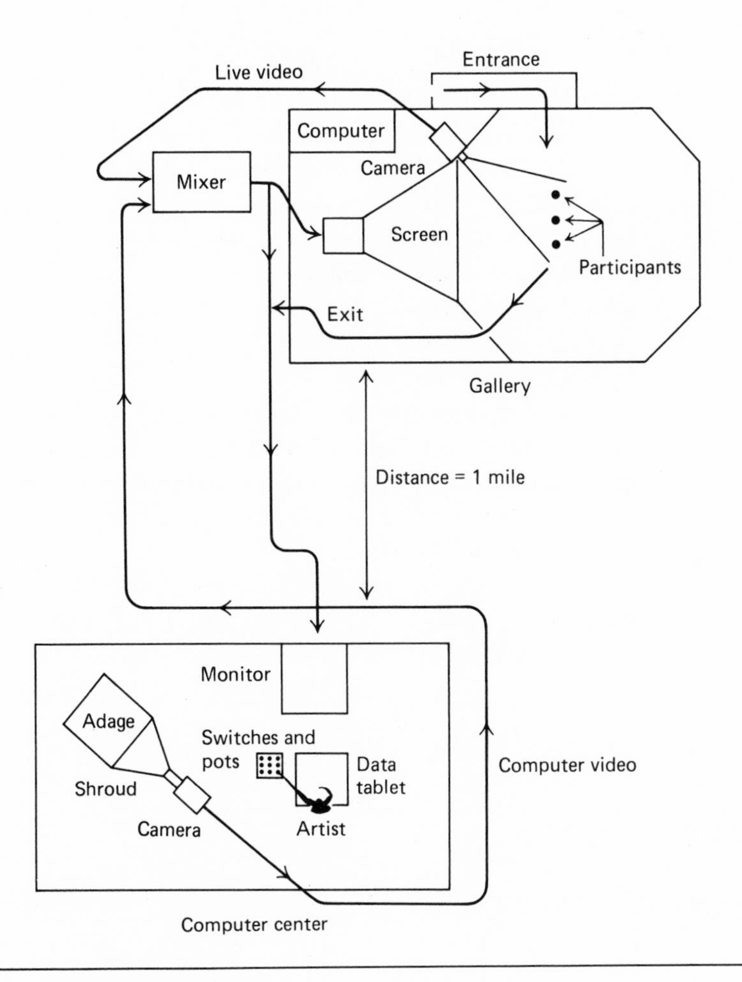

Fig. 3.2 METAPLAY communications

The facilitator could draw on the display screen by moving the pen on the data tablet. By using buttons, switches, knobs, and a keyboard, the facilitator could rapidly modify the pictures generated or alter the mode of drawing itself.

The software provided a variety of techniques for defining picture elements and manipulating them. Any object could be moved about the screen and changed in size. The horizontal and vertical dimensions of an object could be scaled independently of each other. Surprisingly, the visual effect of this manipulation was an apparent rotation of the object in three dimensions. In addition, any element could be repeated up to ten times on the screen. The relative position and size of each repetition could also be controlled (Fig. 3.3). Also, a tail of a fixed number of line segments could be drawn allowing the removal of a segment at one end while another was added at the opposite end. A simple set of transformations controlled by knobs yielded apparent animation of people's outlines. Finally, previously defined images could be recalled or exploded.

While it may seem that the drawing could be done without a computer, the ability to rapidly erase, recall, transform, and animate images required considerable processing and created a far more novel means of expression than a pencil and paper could provide.

These facilities provided a rich repertoire for an odd kind of dialogue. Many approaches were tried, and new ideas cropped up through the duration of the show. Using the draw-

Fig. 3.3 Image element replicated with fixed x,y and scale offsets

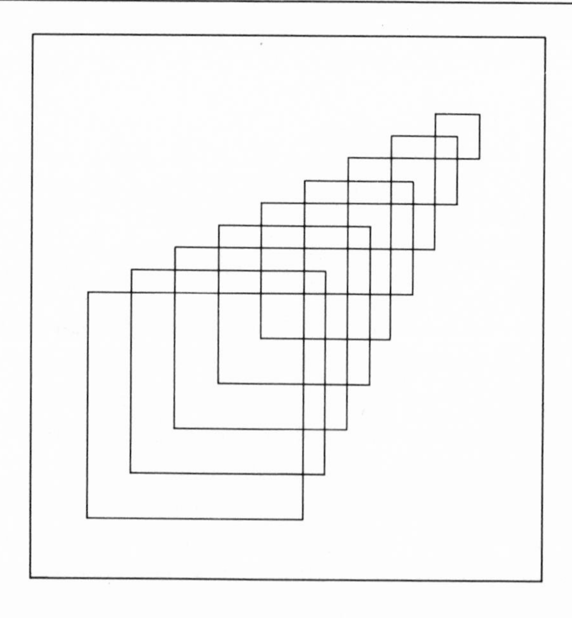

ing tools described above, the facilitator could draw on the participant's images. For instance, a bathtub could be drawn around a seated participant's image in a way which made him or her appear to be seated inside of it (Fig. 3.4). Or, it was possible to draw a graphic door that opened when a participant touched it. Alternatively, the facilitator could communicate directly by writing words or attempt to induce the participants to play a game such as Tic-Tac-Toe. Finally, the act of drawing itself could be played with, as one kind of picture would be smoothly transformed into another.

Fig. 3.4 METAPLAY drawing

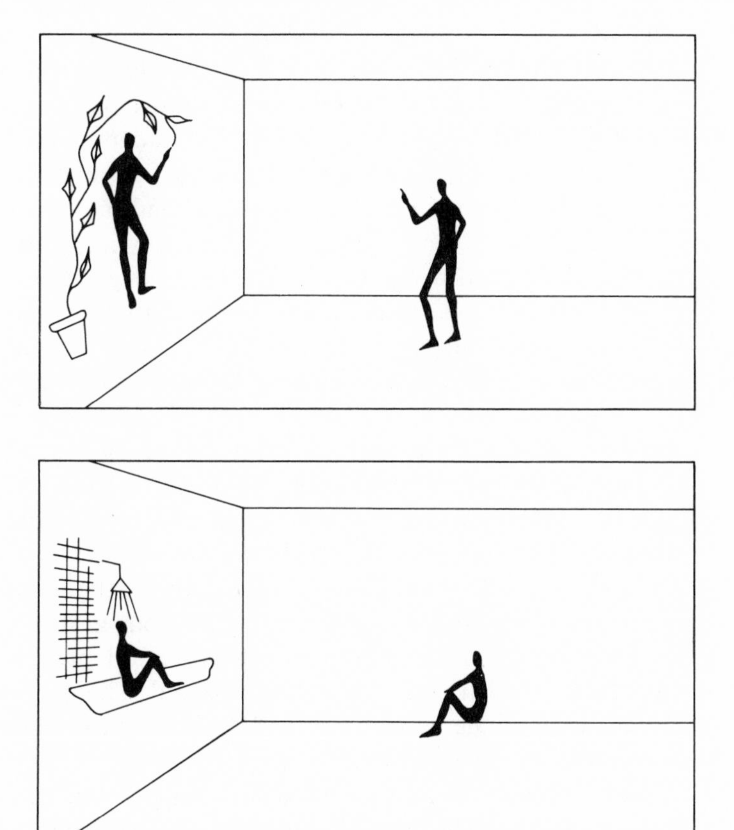

Live Graffiti

One type of interaction derived from the facilitator's ability to draw on the image of the participant. Graffiti features, such as a mustache and beard could be added, or an outline could be drawn around a person and animated so that it appeared to dance to the music in the gallery. The drawing process proceeded with the facilitator trying one idea after another in an effort to involve a particular participant. If that person could not be induced to play, the cameraman in the gallery was directed to focus on a second person. The facilitator thus had control not only of the drawing, but of the camera as well.

One of the most interesting relationships came from our desire to have a way for the people in the Environment to draw. An electronic wand designed for this purpose was not completed. The following serendipitous solution was perhaps preferable. On one of the first days of the show, I was trying to draw on someone's hand. He did not understand what was happening and moved his hand. I erased what I had drawn so far and started over where his hand had moved. Again he moved. This became a game with him moving his hand just before I finished my drawing. The game degenerated to the point where I was simply tracking the path of his hand with the computer line. Thus, by moving his hand he could draw on the screen. This idea became the basis for many interactions.

We tried to preserve the pleasure of our original discovery for each of the people we wanted to involve in this way. After we had played some of the graffiti games with each group, we would focus on a single person. We would busily draw around the image of their hand. The reaction was usually bewilderment. After a minute or so, the increasingly self-conscious person would make a nervous gesture such as scratching his nose. Another minute would pass with the person's hand frozen, while he pondered. Then, a tentative movement of the hand. The line followed. It worked! And he was off, trying to draw. Then others would want to play. Using a finger, the first person would pass the line to someone else's finger, who would carry it to the next. Literally hundreds of interactive vignettes developed within this narrow communication channel.

Drawing by this method was a rough process. Pictures of any but the simplest shapes were unattainable. This was true mainly because of the difficulty of tracking a person's finger as it moved. If it moved slowly there was a chance that something recognizable could be drawn — otherwise not. The data tablet used to put the drawing into the computer was also a low-resolution device, further frustrating real drawing. But neither the facilitator nor the audience was ever concerned by the limitations of the drawings. What excited people was interacting in this peculiar way through a video-human-computer-video communication link spanning a mile.

THE EXPERIENCE

The sequence of events started with a group of six to eight people entering the darkened Environment. The lights were brought up and their projected video images became visible. The typical audience reaction at this point was surprising. Often, faced by the large screen where the only active element was their own image, people would sit down and watch. Large-screen video projection was apparently undreamed of by many of the participants. We therefore allowed at least a minute for them just to appreciate the phenomenon. After the initial awe was overcome, one of the interactions would ensue. These were terminated by the lights dimming and the artist writing "Good-bye," or the equivalent.

COMPUTER-CONTROLLED RESPONSIVE ENVIRONMENT

In the second METAPLAY approach, human-machine interaction was direct rather than mediated by a facilitator. A program monitored the participants' behavior and defined a composed series of feedback relationships using the sound synthesizer and the graphic display.

Hardware

Both the graphic and music responses were based on the positions of people in the room as determined by the floor sensing array. This array consisted of 768 pressure-sensitive switches arranged in a 24 × 32 grid. The large number of sensors compared to the six used in GLOWFLOW reflected our decision that the computer had to know as much as possible about the participants' location and behavior.

The switches were interconnected so that the output from the floor was two voltages representing the coordinates of a person in the room. If the person were standing on several switches, the outputs would be the averages of those switches. One disadvantage of the averaging process was that if a switch became stuck in the "on" position, it would bias the output considerably no matter how the participant moved. Such reliability problems prevented the system from being fully explored.

The position values from the floor were transmitted over telephone lines to the Computer Center, where they were fed into the graphic display computer and used as input for one of a number of different programs. These programs generated visual images on the CRT where they were picked up by a television camera, transmitted to the gallery and projected on the 8′ × 10′ screen. Note that this was the same video system used in the drawing interaction except that no live image was superimposed.

Graphic Interaction

Only one person was allowed in the Environment at a time. This focused participants' attention on the relationship between their behavior and the Environment's response. If several people had been present, their awareness of each other would have competed with their awareness of the displays.

There also would have been confusion about whose behavior was eliciting a particular response.

The participant's movements about the room generated graphic images that were displayed on the projection screen. In most of these, room position corresponded to the position of a cursor on the screen. Thus, one could draw on the screen by walking around the room.

The first few moments of each person's experience were regarded as a training period. A projected dot on the screen moved in correspondence to the person's movements around the floor. When the participant understood this relationship, the computer would move into more complex drawing relationships, each followed by its own surprise or extension. Changes in the mode of drawing or transformation of the image occurred at prescheduled times.

There were a number of simple interactions based on drawing. All of the variations used in the live drawing interaction were available. In addition, the dot following a person's movements could leave a trail of dots marking each of the participant's past positions. Thus it was possible, in a sense, to interact with one's own recent past. If a viewer stood still, the number of dots indicating recently occupied positions decreased until the past caught up to him. Or, the trail of dots might suddenly follow an arbitrary path so the individual's past went on without him.

Responsive Music

The music program, designed by one of my students, was intended to provide a responsive background of sounds for the graphic interactions. The music was composed in terms of rhythmic and tonal complexity. The length of the room corresponded to a continuum of tonal complexity with simple pure tones at one end and progressively more complex sounds towards the other. The breadth of the room controlled the rhythmic complexity with silence on one edge, regular rhythms towards the center, and complex arrhythmias leading to a busy collage of sound at the other edge. Thus, each position in the room caused a certain kind of tonal and rhythmic event. At any given moment, the next sound or rhythm chosen was a random selection from the currently appropriate options which were heavily influenced by whatever the current sounds were. The result was that while there was a deterministic scheme for choosing the levels of complexities, the actual sequence of sounds varied randomly over a period of time. There was no concession to the concept of a "piece" that started when the participant entered and climaxed as the participant left. Rather, the participant's actions had immediate effects, and the general character of the sounds changed slowly but perceptibly as the participant's experience progressed.

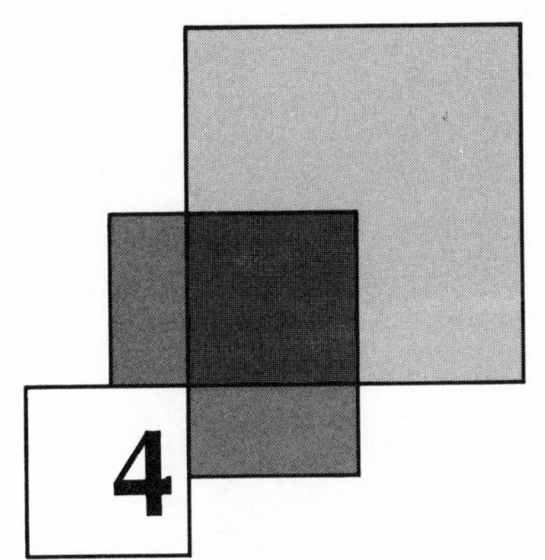

4 PSYCHIC SPACE, HUMAN-MACHINE INTERACTION

CONCEPTION AND DESCRIPTION

PSYCHIC SPACE was exhibited in the Memorial Union Gallery for the month of May 1971. This Environment provided both a richly composed interactive vignette and an instrument for musical and visual expression. Computer-facilitated interaction between people was temporarily set aside so attention could be focused on human-machine interaction.

PSYCHIC SPACE was quite a different experience from both GLOWFLOW and METAPLAY. GLOWFLOW was a group event and METAPLAY included the obtrusive intervention of a facilitator. PSYCHIC SPACE, on the other hand, was the experience of just one person with the Responsive Environment.

Since the reactions were private, the Environment was quieter than it had been in the previous exhibits. Its moods clearly reflected the personality of the individual rather than the spirit of the group. Since the experiences were automated, I observed less participant behavior. Only one attendant stayed with the exhibit, and the attendant's role was minimal —

showing people where to go, answering a few questions, and restarting the computer from time to time. I often left one of my children, who were five and six at the time, in charge. These tots, lurking in the darks of the gallery and apparently competent with computers, set an eerie tone for visitors and became a part of the piece. Computers truly were child's play. I often overheard people in the student union talking about "these incredible kids who were running a computer show in the gallery."

The Environment constructed within the gallery was larger than the space used for METAPLAY. It was large enough to run in, yet still small enough to provide intimacy with all displays (Fig. 4.1). The floor, walls, and ceiling were covered with black polyethylene. One end wall was dominated by a rear projection screen; the other was painted with a phosphorescent pigment. The floor was divided into a sensing grid of six rows of eight 2′ × 4′ modules. Each module was covered with black plastic. Visually this arrangement was very effective, with its suggestion of function and the repetition of modular design (Fig. 4.2). The rear projection screen concealed a television projector while the computer

Fig. 4.1 PSYCHIC SPACE

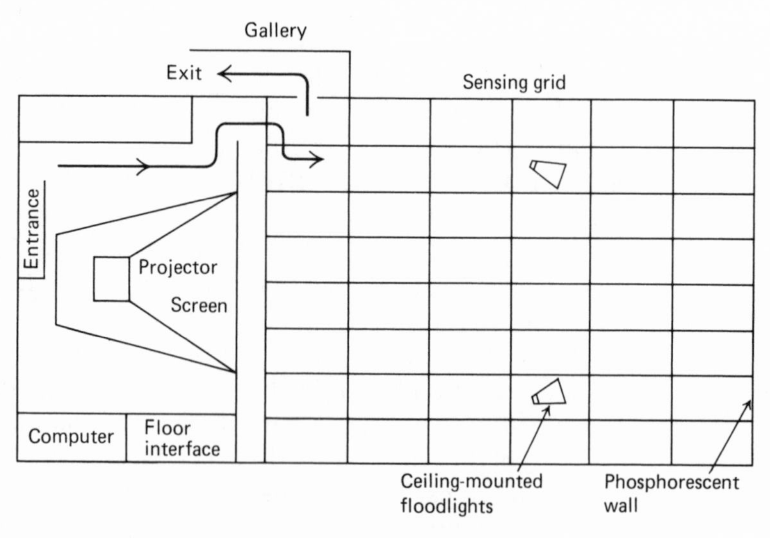

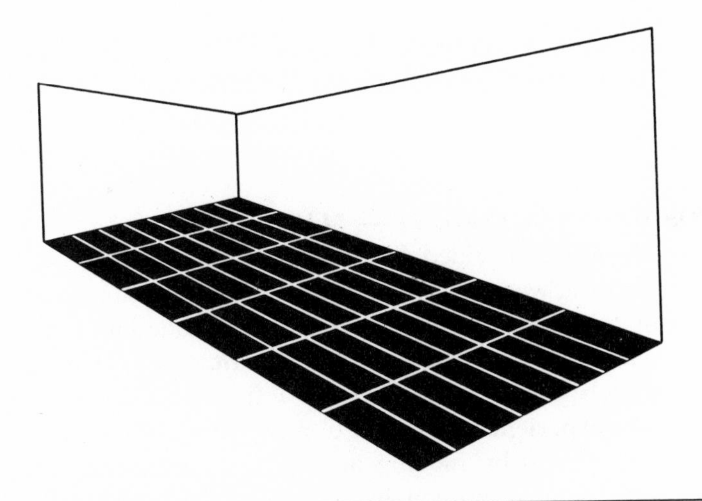

Fig. 4.2 Floor sensing modules in PSYCHIC SPACE

and the circuitry were sequestered in a room next to the projector room. The entryway and exit were designed to limit the amount of light entering the Environment.

HARDWARE

The computer controlling this exhibit was a Digital Equipment Corporation PDP-11 which directed control of all sensing and sound in the gallery. In addition, it communicated with the Adage graphic display computer at the Computer Center. The Adage image was transmitted over video cable to the gallery where it was again rear-projected on the screen.

Position information from the floor was the basis for all of the interactions. The sensing was done by a grid of 400 pressure switches. These switches were constructed in 2′ × 4′ modules. They were electronically independent, enabling the system to discriminate among individuals if several were present. This independence also made it easy for the programs to ignore a faulty switch until its module was replaced or repaired. Since the PDP-11 has a sixteen-bit word, it was natural to read the sixteen switches in each row across the room in parallel. Digital circuitry was then used to scan the twenty-four rows under computer control.

COMPOSED ENVIRONMENT — MAZE

The Maze provided a highly variable interactive vignette. Its purpose was to demonstrate a carefully composed sequence of relations that would result in a coherent experience. It was a finished piece in the artistic sense. It focused completely on one person's interaction with the Environment. The experience consisted of navigating a maze (projected on the screen) by moving around the room. The participant's interest was piqued by the computer's responses to their efforts to walk through the maze.

Maze Hardware

The Maze itself was not programmed on the PDP-11. Rather, it was running on the Adage located a mile away in the Computer Center. The PDP-11 transmitted the participant's floor coordinates across an audio cable to the Adage. The Adage generated the Maze image which was picked up by a TV camera and transmitted via a video cable to the gallery where it was rear-projected to a size of 8′ × 10′ (Fig. 4.3).

The Maze Experience

The Maze Environment was totally dark, with the exception of a symbol that moved when the participant moved. People noticed this very quickly as there was nothing else to see. If you moved to the left, it moved to the left. If you moved right, so did your symbol. If you walked towards the screen, it moved down; and if you walked away from the screen, it moved up.

This initial exploration was a training period that educated participants to the relationship between their movements and the movements of their symbol. After a couple of minutes, a square appeared on the screen. Virtually everyone

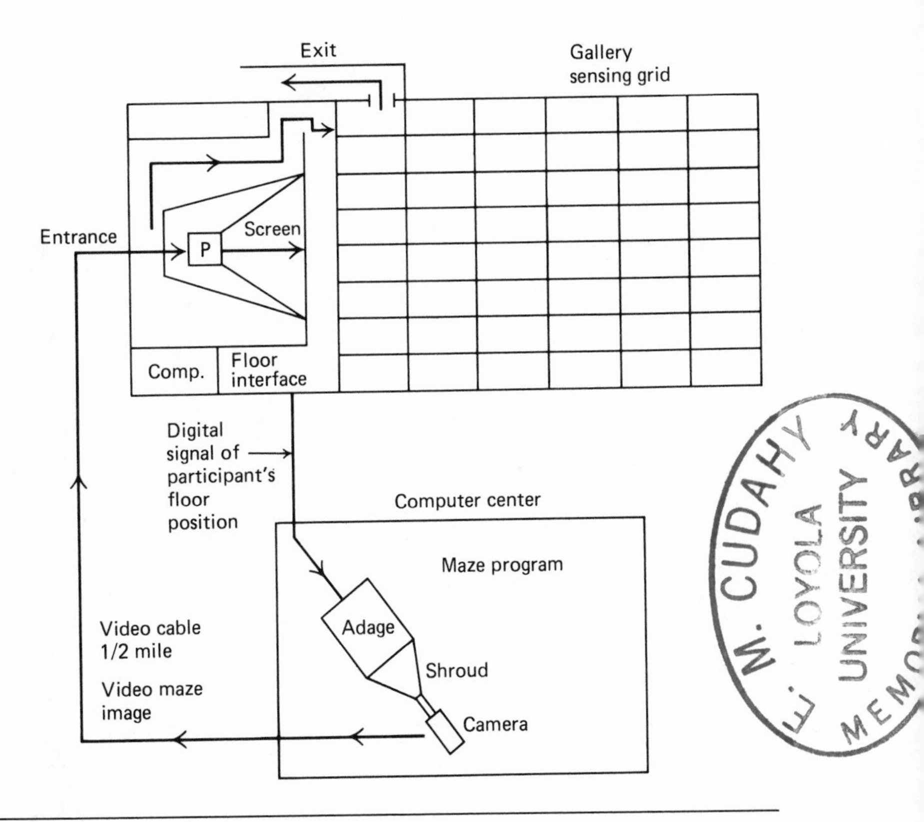

Fig. 4.3 Data and video communication for PSYCHIC SPACE

wondered about the function of this new symbol and walked over to get acquainted. There was little else to do.

When the symbol reached the square, the square disappeared and a maze appeared. Mysteriously, the participant's symbol was at its entrance. In fact, the whole purpose of this introduction of the square was to induce the participant to move to the starting point of the maze. This expedient allowed us to avoid writing a more complicated program that could have generated a maze with an entrance wherever the participant happened to be (Fig. 4.4).

Confronted with the Maze, no one questioned the inevitability of walking through it. (This undoubtedly reflects upon our educational system.) Because the corridors of the

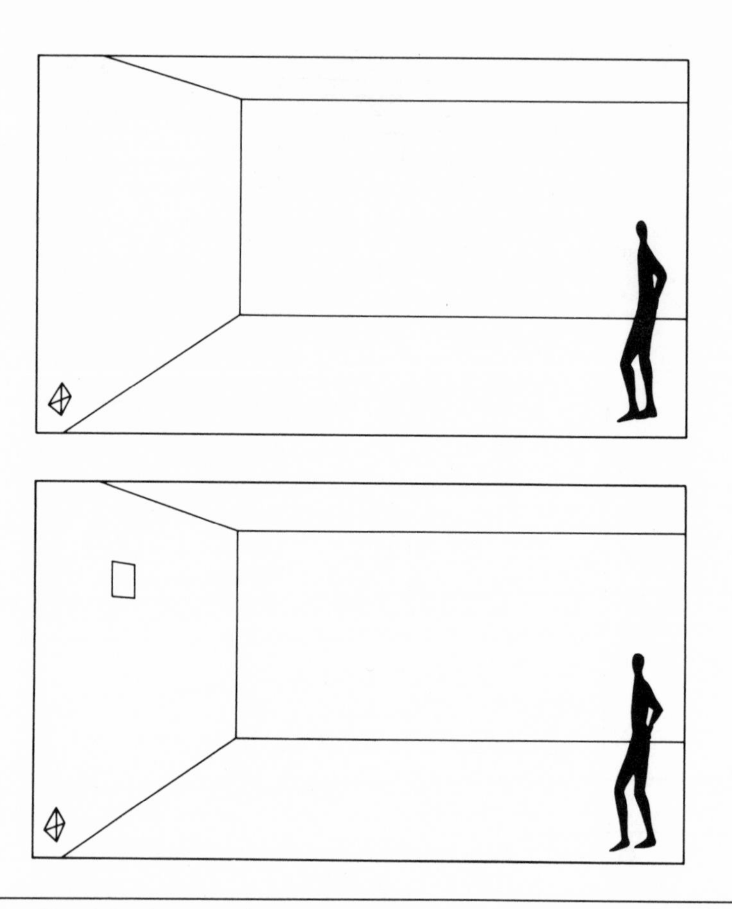

Fig. 4.4a Composed environment—maze

Maze were small relative to the size of the room, it was necessary to take very small steps to navigate it. Participants minced up and down and across the room intently in their efforts to solve the puzzle. Then, after a few minutes of taking tiny steps, participants would realize that there were no physical constraints in the room. In fact, there was nothing to prevent cheating. So, most people began to cheat. Often with great ceremony, they would lift a foot and place it on the other side of a boundary. Now the real Maze began. The cheating response had been anticipated and a large number of ways to thwart it had been composed.

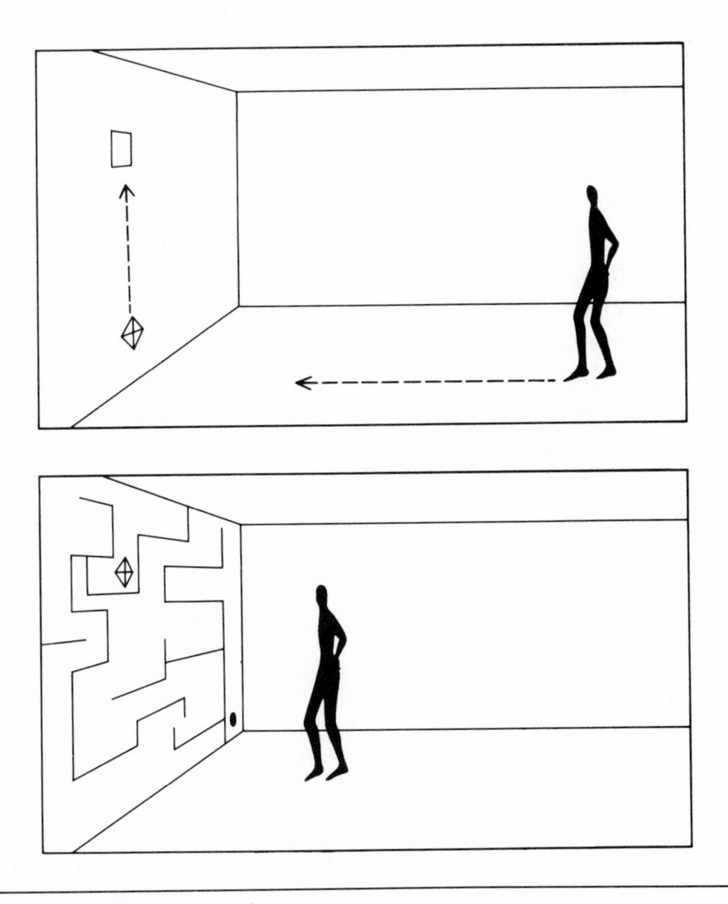

Fig. 4.4b Composed environment—maze

The first time a participant tried to cross a boundary line, it would stretch elastically, keeping the participant's symbol enclosed. Later, the line would disappear, removing a barrier, but the rest of the Maze would change simultaneously, eliminating the apparent advantage. Next, the person's symbol would split in half at the violated boundary, with one half remaining stationary while the other half, the alter ego, continued to follow movement. However, no progress could be made until the halves of the symbol were reunited at the violated boundary. At another point, the whole Maze would move so that the participant's symbol appeared to push it off the screen. In all, there were about twenty different ways the computer would retaliate if the participant cheated.

Even when the participant was moving legally, there were changes in the program contingent upon position in the room. Several times, as the goal was approached, the Maze changed to prevent immediate success. Or, the relationship between the floor and the Maze was altered so that movements that once resulted in vertical motion now resulted in horizontal motion. Alternatively, the Maze could be moved while the symbol representing the participant was kept stationary.

Ultimately, success was prohibited. When reaching the goal seemed imminent, additional boundaries appeared in front of and behind the symbol, boxing it in. At this point, the Maze slowly shrank to nothing. While the goal could not be reached, the composed frustration made the route interesting.

The Maze experience induced a unique set of feelings. The video display space created a sense of detachment enhanced by the displaced feedback; movement on the horizontal plane of the floor translated onto the vertical plane of the screen. The popular stereotype of dehumanizing technology seemed fulfilled. However, the Maze idea was engaging, and people became involved willingly. The lack of any other sensation focused attention completely on this interaction. As the experience progressed, their perception of the Maze changed. From the initial impression that it was a problem to solve, they moved to the realization that the Maze was a vehicle for whimsy, playing with the concept of a maze and poking fun at their compulsion to walk through it.

THE ENVIRONMENT AS A MUSICAL INSTRUMENT

The floor in PSYCHIC SPACE had a dual role. In addition to reporting the location of a person walking the Maze, it was also used as the keyboard of a musical instrument that the participants could play by moving around the room.

When a person entered the room, the computer automatically responded to their footsteps with electronic sound. We experimented with a number of different schemes for actually generating the sounds based on an analysis of people's footsteps. In sampling the floor sixty times per second we discovered that a single footstep consisted of as many as four discrete events: lifting the heel, lifting the toe, putting the heel down, and putting the ball of the foot down. The first two were dubbed the "unfootstep." We could respond to each footstep or unfootstep as it occurred, or we could respond to the person's average position. A number of response schemes were tried, but the most pleasing was to start each tone only when a new switch was stepped on and then to terminate it on the next "unfootstep." Thus, it was possible to get silence by jumping, or by lifting one foot, or by putting both feet on the same switch.

The participants' typical reaction to the sounds was instant understanding, followed by a rapidfire sequence of steps, jumps, and rolls. This phase was followed by a slower and more thoughtful exploration of the Environment in which more subtle and interesting relationships could be developed. In the second phase, the participant discovered that the room was organized with high notes at one end and low notes at the other. After a while, the keyboard was abruptly rotated ninety degrees, challenging what the participant had learned.

After a longer period of time an additional feature came into play. If the computer discovered that a person's behavior was characterized by a series of steps punctuated by relatively long pauses, it would use the pause to establish a new kind of relationship. The sequence of steps was responded to with a series of notes as before; however, during the pause, the computer would repeat these notes again. If the person remained still during the pause, the computer assumed that the relationship was understood. The next sequence of steps was echoed at a noticeably higher pitch. Subsequent sequences were repeated several times with variations each time. This interaction was experimental and extremely difficult to introduce using feedback rather than explicit instructions. The goal was a human-machine dialogue resembling the guitar duel in the film "Deliverance." This dialogue was an interesting experience, especially when it was discovered by the participant without prior instruction.

While the sounds were not complex, a variety of styles were observed. The differences in the sounds reflected the differing natures of the participants. Some people seemed to want to cause as many different sounds as possible. Others would audibly tire and silently rest. Others would try to pick out a tune, which was difficult. Occasionally, quite distinctive sounds would emerge, prompting me to investigate their source. A few times, very monotonous sequences of tones signalled someone who was totally unimpressed by the sound responses, or perhaps deaf to them. They just walked as they would anywhere else. Thus, there was no intelligent relationship between one sound and the next. One time I heard an extremely atypical pattern of single notes, separated by long pauses. I found an Oriental monk. He would take a step, stop to ponder its consequences and only after some delay, take another. He professed to be quite moved by his experience.

On another occasion, extremely pleasant bubbling sounds greeted me as I returned to the gallery. I learned from those who had been looking after the exhibit that a very attractive couple was dancing together. When they came out we asked for an encore.

PHOSPHORESCENT SHADOWS

In one composed interaction, the musical responses gradually led the participants towards the phosphorescent panels which were at one end of the Environment. These could be illuminated by flood lights under computer control. In the darkened Environment people were exceptionally quick to respond to the auditory cues we used to guide them. The cues were sounds triggered only by movement in the desired direction. An ultrasonic motion detector, similar to those used in home burglar alarms, was used in conjunction with the floor to determine whether the person was standing still in front of the panels. Since the phosphors retain light, illuminating the panels caused the whole wall to glow except where the participant's body blocked the light. When the flood lights were turned off, a detailed life-sized silhouette was left frozen on the wall. We found that it was useful to warn participants when the light was about to flash to allow them time to prepare an interesting pose. This was accomplished by preceding the light flash with a sequence of ascending pitches. At the end of the sequence, the light would come on for about one second. After the light went off, time was provided to admire the shadow before the process was repeated. This interaction was the simplest; however, it served its purpose as backup in case the more complex systems temporarily failed.

In an Environment with a much wider array of response displays, this system would be used only with people who were in the required location. Other people would receive other responses from the Environment. For example, if the computer were to see someone wildly flailing his or her arms to get an audio response from an ultrasonic system, it would be effective to stop the sound abruptly and flash the light, capturing the person's gesture. The phosphors would become a dramatic surprise for a few people rather than an anticipated event for everyone.

GLOWMOTION

While the phosphorescent panels were very popular as part of an aesthetic Responsive Environment, they can also stand alone as a form of entertainment. With this in mind, we took a show called GLOWMOTION to county fairs in rural Wisconsin in the summer of 1972. In this exhibit, an extremely bright flashtube was used instead of the flood lights. The flash produced was so brief that it was possible to capture a jumping person's shadow in midair. The ability to freeze the image as the person moves is important: it allows the effect to be more smoothly integrated with the Responsive Environment. Today we can duplicate the GLOWMOTION effect by freezing a video image.

CONCLUSIONS

During PSYCHIC SPACE we got a better feeling for the mechanics of putting on an exhibit, of getting people in and out and of minimizing the amount of verbal instruction required. At the same time, there were some misgivings about the concept of this type of exhibit which inevitably attracted more people than it could handle. Even without any publicity, METAPLAY and PSYCHIC SPACE always had long lines of people waiting to participate. As a result, we had to gear the experience to handle the crowds, making it shorter and allowing more people in at one time. Since the Maze experience lasted about fifteen minutes, it had to be used sparingly. But this compromised the whole point of the work, personal interaction with an Environment. So, I became interested in creating an ongoing facility in which the conceptual, compositional, and technical aspects of Responsive Environments could evolve.

5 THE RESPONSIVE ENVIRONMENT, A NEW AESTHETIC MEDIUM

SATURATION OF THE SENSES

Responsive Environments are particularly timely because of the current state of the arts. For some time artists have lamented the diminishing effectiveness of their traditional tools. It is often said that painting is dead. Art historian Jack Burnham suggests that art itself is dead. To these obituaries, I would like to add my own somewhat hyperbolic statement and then examine to what extent it is true.

The visual is dead! We are constantly bombarded with visual images on TV, in magazines and movies. Not only is there a quantity of visual images, there is undeniably quality as well. Most of the images we see are carefully crafted for maximum effect and many are beautiful. The result is that we cannot take in all we see. Numbed by the onslaught of visual information, insulated by categories and filters, vision, our most heavily trafficked sense, is no longer capable of reacting to paintings or graphics as art. Sending a message through vision alone is like sending it through channels; you can be sure that it will be processed correctly, but also that it will be treated as routine. To touch a person today, you have to slip past defenses and involve him in an unfamiliar way.

BEYOND INTERPRETATION

Oddly, art history, art criticism, and art appreciation have become deterrents to experiencing art. Repeatedly, people leaving GLOWFLOW, METAPLAY, and PSYCHIC SPACE said "I really liked it, but what did it mean?" For some reason they felt that what had happened should be immediately reduced to words. In fact, there is a tendency to accept events in terms of the words that will be used to describe them. Therefore, there is a place for a medium that can resist interpretation. The Responsive Environment can take steps to individualize the responses and to thwart analysis. If each person has a different experience, there will be less pressure to arrive at the "right" interpretation. Since each person moves about the space somewhat differently, each will receive different feedback, even if the controlling program is exactly the same. If there are many programs alternating control of the Environment, each participant's adventure will be unique. Thus, two people can exchange experiences, but since they have had no common experience they cannot analyze it to death.

ACTIVE VERSUS PASSIVE ART

There is another way that Responsive Environments answer a cultural need. All of our traditional art forms have one thing in common: they assume a passive audience. Passivity was appropriate when men toiled physically. However, after centuries of effort we have all but eliminated the necessity for physical exertion. Ironically, since our bodies require a certain amount of exercise for health, we face a new problem — how to reintroduce labor into our lives. Sports do this for some, but we also need new forms of art and entertainment that will involve our bodies rather than deny them.

RESPONSE IS THE MEDIUM

The Environments described suggest a new art medium based on a commitment to real-time interaction between people and machines. The medium is comprised of sensing, display, and control systems. It accepts inputs from or about a participant and then makes responses (outputs) in a way they can recognize as corresponding to their behavior. The relationship between inputs and outputs is variable, allowing the artist to intervene between the participant's action and the results perceived. Thus, for example, a participant's physical movement can cause sounds, or their voice can be used to navigate a computer-defined visual space. It is the composition of the relationships between action and response that is important. The beauty of the visual and aural response is secondary. Response is the medium!

In principle, the Environment can respond to a participant's position in the room, voice volume or pitch, position relative to prior position, or the time elapsed since the last movement. It can also respond to every *n*th movement, the rate of movement, posture, height, colors of clothing, or time elapsed since the person entered the room. If there are several people in the room, the Environment can respond to the distance separating them, the average of their positions, or the computer's ability to perceive them as separate entities when they are very close together.

In more complex interactions like the Maze, the computer can create a context within which the interaction occurs. This context is an artificial reality within which the artist has complete control of the laws of cause and effect. Thus the actions perceived by the hardware sensors are tested for significance within the current context. The computer asks if the person has crossed the boundary in the Maze or has touched the image of a particular object. At a higher level, the machine can learn about the individual and judge from its past experience with similar individuals just which responses would be most effective.

Currently, these systems are constrained by the total inability of computers to make certain very useful, and for human beings, very simple perceptual judgments, such as whether a given individual is a man or a woman or is young or old. The perceptual system will define the limits of meaningful interaction, for the Environment cannot respond to what it cannot perceive.

As mentioned before, the slickness of the displays is not as important in this medium as it would be if the form were conceived as solely visual or auditory. In fact, it may be desirable to have displays that are not beautiful in any sense, for that would distract from the central theme: the relationship established between the observer and the Environment. Artists are fully capable of producing effective displays in a number of media. This fact is well known and to duplicate it produces nothing new. What is unknown and remains to be tested is the validity of a responsive aesthetic.

However, it is necessary that the output media be capable of displaying intelligent, or at least composed reactions, so that the participant knows which of their actions provoked it and what the relationship of the response is to the participant's action. The purpose of the displays is to communicate the relationships that the Environment is trying to establish. They must be capable of great variation and fine control. The response can be expressed in light, sound, mechanical movement, or through any means that can be perceived.

CONTROL AND COMPOSITION

The control system includes hardware and software control of all inputs and outputs as well as processing for decisions that are programmed by the artist, who must balance the desire for interesting relationships against the commitment to respond in real time. The simplest responses are little more than direct feedback of a participant's behavior, allowing the Environment to show off its perceptual system. But far more sophisticated results are possible. In fact, a given aggregation of hardware sensors, displays, and processors can be viewed as an instrument that can be programmed by artists with differing sensitivities to create completely different experiences. A Responsive Environment can be thought of in the following ways:

1. An entity that engages participants in a dialogue. The Environment expresses itself through light and sound while the participant communicates with physical action. Since the experience is an encounter between individuals (human and machine), it might legitimately include greetings, introductions, and farewells — all in an abstract rather than literal way. The artist's task in this case is similar to a writer's — to make the Environment into an interesting character by giving it a distinctive personality.

2. A personal amplifier. In Dorothy's initial encounter with the Wizard of Oz, the Wizard uses technology to enhance his ability to interact with those around him. The Environment can be used in the same way. The live drawing interaction in METAPLAY is an example of this approach.

3. A multidimensional space which the participant can explore, where each subspace has unique response relationships. This space could be inhabited by artificial organisms defined either visually or with sound. These creatures can interact with the participants as they move about the room.

4. An amplifier of physical position in a real or artificially generated space. Movements around the Environment would result in much larger apparent movements in the visually represented space. A graphic display computer can be used to generate a perspective view of a modelled space as it would appear if the participant were within it. Movements in the room would result in changes in the display: by moving only five feet within the Environment, the participant would appear to have moved fifty feet in the display. The rules of the modelled space can be totally arbitrary or physically impossible, e.g., a space where objects recede when one approaches them.

5. An instrument that participants play by moving about the space. In PSYCHIC SPACE the floor was used as the keyboard of a simple musical instrument.

6. A means of turning the participant's body into an instrument. The person's physical posture would be determined from a digitized video image, and the orientation of the limbs would be used to control lights and sounds.

7. *A game between the computer and the participant.* This variation is really a far more involving extension of the pinball machine or the video game, already the most commercially successful interactive environments.

8. *An experiential parable where the theme is illustrated by the things that happen to the protagonist — the participant.* Viewed from this perspective, the Maze in PSYCHIC SPACE becomes pregnant with meaning. It was impossible to succeed, to actually solve the Maze. This could be a frustrating experience if one were trying to reach that goal. If, on the other hand, the participants maintained an active curiosity about how the Maze would thwart them next, the experience proved amusing and thought provoking. Such poetic composition of experience is one of the most promising lines of development to be pursued within Responsive Environments.

IMPLICATIONS OF THE ART FORM

For artists, the Responsive Environment augurs new relationships with their audience and their art. The artist operates at a metalevel. The participant provides the direct performance of the experience. The Environmental hardware is the instrument. The computer acts much as an orchestra conductor, controlling the broad relationships while the artist provides the score to which both performer and conductor are bound. This relationship might be a familiar one for musical composers, although even they are accustomed to being able to recognize one of their pieces, no matter who is interpreting it. But the artist's responsibilities here become even broader than those of a composer who typically defines a detailed sequence of events. The interactive artist is composing a network of possibilities, many of which will not be realized for a given participant who fails to take the particular path along which they lie.

Since the artist is not dedicated to the idea that the entire piece be experienced, it is possible to deal with contingencies, by trying different approaches, different ways of eliciting participation. The artist can take into account the differences among people.

In the past, art has often been a one-shot, hit-or-miss proposition. Paintings hanging in rows in galleries or museums have the same problem as boxes of detergent at the supermarket. Somehow, each must distinguish itself from its competitors if it is to attract the viewer's attention. Perhaps in response to this need, paintings have become very large. However, even after the viewer has chosen a painting to look at, they can be easily distracted by the painting next to it.

An interactive Environment deals with the problems of getting and maintaining attention in novel ways. First, it refuses to compete with other pieces. It offers nothing to the casual observer. A commitment is required before anything is revealed. It responds only when the participant has entered its space. Thereafter, the Environment can judge the participant's continuing involvement and modify its own behavior if the participant's interest begins to wane. It can learn to improve its performance, responding not only to the moment but also to the entire history of its experience with other participants. The piece becomes an aesthetic entity whose behavior will mature through experience and which may take paths unanticipated by the artist. Indeed, one of the strong motivations guiding this work is the desire to create works that surprise the creator.

Participants are confronted with a completely new kind of experience. They are stripped of informed expectations and forced to deal with the moment in its own terms. Participants are actively involved, discovering that their limbs have been given new meaning and that they can express themselves in new ways. The experience achieved will be unique to each person's movements and may go beyond the intentions of the artist or what the artist had thought were the possibilities of the piece.

Using the Responsive Environment as a vehicle for implicit content also raises the question of what is communicated by the medium itself and what it can be used to communicate.[1] First, the medium presents some unavoidable facts about current technology. For better or worse our technology is going to perceive us. It will communicate with us. The relationship is going to get cozier and more intimate as time passes. The Responsive Environment introduces some of the most up-to-date technology in a way that makes its implications palpable. Since it is neither technical explanation nor histrionics about dehumanizing technology, the experience can serve as an early warning system for people who seek to know what they may be called upon to adapt to.

TECHNOLOGY FOR FUN

More important than the specific knowledge a person may gain about technology is the attitude that is conveyed. The Responsive Environment is technology for fun. Americans are incredibly attuned to the idea that the sole purpose of technology is to solve problems. We seem unable to grasp that only by completely integrating our technology with the whole of our lives can we understand its implications sufficiently to use it with confidence to solve our problems. Consequently, with the recent and probably temporary exception of video games, we buy entertainment equipment almost exclusively from other countries which are better able to see the implications of our inventions in terms of day-to-day life.

In addition, these Environments illustrate ways that technology can be personalized and humanized. It is possible to program the Environment so that each person has a dramatically different experience not only because they act differently but because the relationships that govern the interaction are different.

Finally, in an exciting and frightening way, the Environments dramatize the extent to which we are savages in the world which our technology creates. The layman has extremely little ability to define the limits of what is possible with current technology and so will accept all sorts of cues as representing relationships which in fact do not exist. The constant birth of such superstitions indicates how much we have already accomplished in mastering our natural environment and how difficult the initial discoveries must have been.

This medium also makes a serious indictment of our current style. It offers nothing to the passive audience. A passive individual can enter, and ignoring the invitation to become involved, leave, having experienced nothing. While in some programs the Environment may be willing to cajole the participant into a conversation, in others it might choose not to bother. The way the Environment treats its participants will reflect the attitudes of the artist.

AESTHETIC ISSUES

Is the Responsive Environment art? Definitions of art are never very satisfactory. A particular type of work may only be considered art for a brief period. Is the Responsive Environment just a passing social statement, an aesthetic one-liner, or does it contain the seeds of a new branch of aesthetic endeavor? If so, which dimensions of the Responsive Environment have the greatest aesthetic potential?

Interactive art shares concerns with existing art forms. First, the Responsive Environment, equipped with a video display, is about the human image, one of the most consistent features of Western art. The focus of the interaction in METAPLAY and VIDEOPLACE (discussed in Chapter 9) is how the participant's image is displayed and what happens to it. People have always been fascinated with their own images. An art form that challenges a person's self-perceptions will continue to be of interest. Responsive art is fundamentally conceptual. Each action elicits a response from the Environment. Since the response is not the expected everyday result of that action, participants are forced to think. They must constantly conceptualize a theory that explains the experience they are having. As the relationships change, participants must update their theory or create a metatheory that ties the pattern of changes together. The Environment becomes a symbolic space where another kind of reality is sketched.

In addition, an artist's actions have become an accepted subject of art. Willem de Kooning paints in a way that reveals the physical act of painting. Jackson Pollock made his physical acts the subject of his painting. In the Responsive Environment, the artist is again observing and commenting on movement. However, in this case, the action of the participant, rather than that of the artist, is the subject of the work.

While making movement a subject of the Responsive Environment relates it to traditional art, switching the focus from artist to participant constitutes a radical departure. True, one can move about and admire a sculpture and to a minimal degree interact physically with a painting. However, the Responsive Environment has closer ties with the Happenings of the early 1960s in its attempt to involve the audience, than it does with these more conventional modes of art. When participation becomes the subject of the aesthetic work, the viewer's critical faculties are impaired. They are no longer interacting solely with the finished work of an artist. The viewer's actions complete the piece. Thus, within the framework of the artist's Environment, the participants also become creators.

From this perspective, participation takes on a different light. Usually, artists are allowed to act, and the audience is not. Dancers are strenuously involved in their work and feel aesthetic pleasure from their own performance. It can even be argued that passive appreciation of any art form is enhanced by the experience of having tried to create. The graceful movements of a dancer take on new meaning to a member of the audience who has tried to dance, just as a painter can identify with the sensuality of another artist's painting experience as revealed by the brushstrokes.

This proprioceptive sense can be more directly addressed in the Responsive Environment. Awareness of one's body is a vital part of the medium. When a person finds that a bend in the elbow has one effect and tilting the head has another, the person discovers a new way of relating to his body. The body becomes a set of transformations that operate on reality. While this is always true, changing the mundane relationships drives the point home. Similarly, as the relationship between the person's actions and their effect upon the Environment changes, the participant is led to ponder his relationship to reality itself. In the real world, the relationship between immediate cause and effect is usually predictable. A consistent set of physical laws mediate our everyday experience. While there are surprises, these usually have social origins or occur because we are unaware of the presence of particular objects and forces, not because the laws governing their behavior are unexpected. Thus, by creating an interaction in which the laws controlling the relationship between action and response are composed rather than consistent, interactive art offers a way to comment on experience itself at a philosophical level.

Just as music addresses the intellectual machinery with which we understand sounds, particularly speech sounds, interactive art can seek the primitive concepts through which we apprehend physical reality. The Environmental experience can be composed in terms of our abstract sense of spaces and objects and the expectations we have for the effect of our actions on the world. If the pattern of confirmed and broken expectations has a coherent and satisfying structure, the result should be an aesthetic experience. As the underlying mechanisms become understood, it should be possible to create a great variety of compositions that explore this part of our makeup.

The Responsive Environment has the potential to endure as an art form. It shares traditional art's concerns about perception, the human image, and the representation of human experience. However, the interactive version of beauty will stimulate conceptual insight as well as perceptual pleasure.

DESIGN CONSIDERATIONS

In designing these experiences there are a host of concerns. The optimum number of people who are to participate at a given time is a crucial issue. If there are several people, the relations among them start competing with those relationships defined by the hardware. If response is the medium, it is clearest when one person is alone with the Environment. If there are a number of people in the space, there must be a way for each to associate their actions with a corresponding response by the Environment; otherwise, the responsiveness becomes meaningless.

One concern is the role of randomness in the Responsive Environment. The Environment is related to the Happening in that the participant is a source of unpredictability. However, it is very important that the responses the participant receives be predictable. If they are not, the participant will lose all sense of control over and all interest in them. It is legitimate to deviate from an expected pattern or to create a new pattern, but there must be enough consistency for the participant to recognize the change. On the other hand, with both sound and graphic patterns it is possible for the artist to describe the general texture desired and then to use random functions to generate the low-level details. The advantage of this approach is that it is easy to create a large number of variations around a well-defined single theme.

The physical structure is another important issue. The Environments that have been implemented thus far were housed in darkened, empty, rectangular rooms. If a shaped space were substituted, the compositions would reflect ideas suggested by the shape — shape would determine content. The empty rectangle has the advantage of being so familiar that physical space is eliminated as a concern and response is the only focus. The walled space is much like the frame for a picture or a pedestal for a sculpture; it separates the composition from the rest of the world. By blocking all distraction, it allows the artist to control the experience. In a completely dark Environment, the participant can perceive only what the artist chooses to show. The darkness also helps free people from inhibitions, making them less self-conscious and more playful.

There are situations where it would be desirable to bring responsiveness into the everyday environment, particularly to create a sort of Happening or an active space that was delineated only by its effects. But in general, this new medium requires people's focused attention so that it cannot be relegated to the status of responsive Muzak.

Finally, I think it is desirable to avoid giving explicit instructions. New relationships should be introduced by a natural sequence of events and understood by exploration rather than explanation. Since the discovery process must proceed at its own pace, this requirement is not consistent with galleries that must often allow large numbers of people through an exhibit. However, reliance on explicit explanation is not worth the time saved, for, unexpectedly, the most obtrusive presence is that of an authoritative human. This reminds us that dehumanization is usually the result of human decisions, not of technology itself.

One situation where explanation must be avoided is at the end of an experience. The problem of ending a piece exists in any medium. The ending should be self-explanatory and consistent with the rest of the piece. The most elegant solution so far has been to respond only to motion toward the exit. This strategy has invariably resulted in people moving in the desired direction. The experience is over when the person finds that they are no longer in the Environment.

The Responsive Environment opens a new dimension for the arts just at the moment when the power of existing forms seems to be on the decline. The focus on live interaction allows the artist to compose a rich variety of alternatives rather than a set of final decisions. The freedom from finality allows a piece to grow, allows the computer to learn. The artist is its sire and each participant a contributor. The responses the computer provides will be as complex as the technology allows. The interactions will take participants into an exploration of their own senses and their own mental processes. Whether the result is high art or mere entertainment does not matter. The Responsive Environment can be used for both purposes. Both are important and any entertainment that is sufficiently inventive and well-crafted, that expresses fundamental human concerns, is a work of art.

Ultimately, however, the Responsive Environment is more than an art medium; it is a whole new realm of human experience.

NOTE

1. H. Marshall McLuhan, *The Medium Is the Massage* (Random House, 1967).

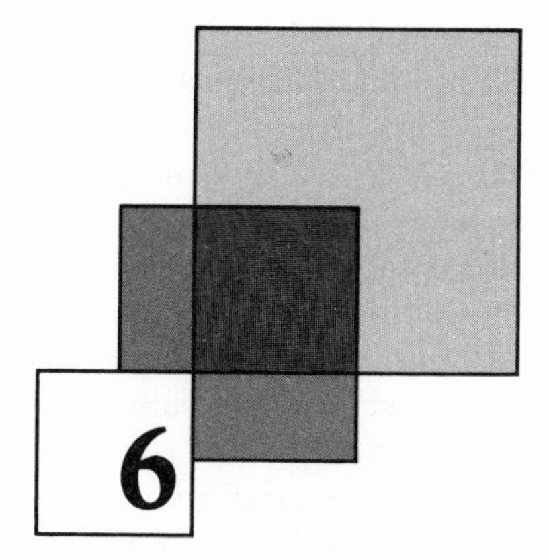

PERCEPTUAL SYSTEMS

DEFINITION AND IMPORTANCE

By definition, a Responsive Environment responds to human behavior. Its ability to respond effectively depends on the quality of the computer's sensors and its ability to interpret what it perceives. The computer receives hardware inputs from either of two sources: obvious devices such as buttons or levers operated by the participant or artist, or hidden sensors used by the computer to perceive the state of the Environment. These inputs, their associated circuitry, and the interpreting software comprise what will hereafter be referred to as the sensory or perceptual system.

This system may yield either static or dynamic information about the participants. Static information includes attributes such as height, weight, or the color of clothing, which the participant cannot change or control during the experience. In fact, the computer can respond to a static attribute, but such a response is unlikely to be meaningful to the participant who cannot identify its cause or control the effect. Dynamic information, like position in the room, rate and direction of movement, and pitch or volume of the voice can be controlled. Thus, while static information may prove useful for identifying or discriminating among individuals when several are present, it is the dynamic information that is used to form the basis of the interactions.

The apprehension of events in the Environment is a two-step process: first, the conversion of sensory information such as sounds, light intensity, and pressure into signals compatible with computer circuitry; and second, the software interpretation of this data to identify the events of interest.

The inputs define the limits of meaningful interaction between the computer and the participants. The Environment's perceptual system determines what the computer knows and thus what it can respond to. If the computer is only aware of the pressure of the participant's feet, arm waving will be invisible and irrelevant to the interaction. Behavior that evokes no response from the Environment is less likely to recur. After the first few minutes, a participant will typically behave only in ways that the computer can perceive. Thus, the mode of sensing can give importance to a certain kind of behavior. Switching to a different mode of sensing creates a new focus and constitutes an important aesthetic event because it changes the nature of the ensuing interaction.

WHAT SHOULD BE SENSED?

Since the goal of the Environment is to involve the participants physically in a relationship they can understand, it should be possible for participants to recognize which aspects of their behavior are eliciting responses. The first experience with feedback usually leads a new participant into a rapidfire sequence of large movements. Initially, the computer should concentrate on the perception of these gross physical movements and postures because the participant will only be aware of this level of behavior. After this phase has run its course, it is possible to involve the participant in more subtle relationships involving small movements, such as nods, shrugs, or even facial expressions.

The design of the perceptual system requires a trade-off between the need for the computer to know as much as possible about the state of the Environment and its commitment to respond in real time. The first responsive sculptures suffered from limited sensory systems, usually consisting of just a few discrete sensors such as photocells or proximity switches providing little information about what was going on in their vicinity.[1] Such sculpture could offer no more than the most simple, reflexive responses. GLOWFLOW, with input from only six pressure pads, was also unable to adequately sense what was happening within its walls.

In order to behave more intelligently, the processor must have a wealth of information. It must be sensitive to small movements, able to identify individuals if several are present and capable of keeping track of those individuals as they move around. On the other hand, the sensory system should not provide too much information. Unneeded detail will require extra processing which might be better used interpreting more limited input or enriching the output responses. The early robot "Shakey," developed at Stanford Research Institute, received so much visual information and devoted so much time to processing that it was unable to behave in real time, requiring minutes to react to a single percept.[2] When the sensor provides more information than the computer can handle, the computer must focus on small parts of it, to keep the amount of data being considered consistent with the real time nature of the medium.

If more than one person is allowed in the Environment at a time, it is desirable for the computer to be able to identify and respond uniquely to each person's movements enabling individuals to recognize the results of their own behavior. However, the ability to identify individuals does not preclude responding to the group as a whole, or to relationships among members of the group.

Sensory hardware should facilitate efficient processing of its inputs by the computer. Arranging the sensors in a rigid format allows the computer to preserve relationships between the inputs of different sensors, especially if that external format can be realized directly in the internal computer model, as was the case in PSYCHIC SPACE where the grid of sensors was represented by an array of twenty-four sixteen-bit words in the computer. This correspondence made it extremely convenient to respond to position and changes in position.

SHOULD SENSORS BE HIDDEN?

The choice of input method includes one purely aesthetic consideration: should the participant be aware of exactly how they are being perceived by the machine? If the answer is affirmative, there are several degrees of obviousness. Most blatant would be explicit controls such as buttons or levers placed around the room, or devices that the participant carries or wears. A less obtrusive but still obvious method would be a number of laser beams scanning the space for people who who will obstruct the light and reflect it back to sensors mounted on the lasers.

There are many possible criteria for making this decision depending on the goals of each particular interaction. In certain instances, a magic wand, a Hobbit's ring, or a blinking cap might add to the experience. My bias is away from explicit controls and encumbering devices because they shift focus from the movements of the body onto an external instrument.

SYSTEMS OPTIONS

Detecting the Participant's Position in the Room

The location of the participant can be determined in a number of ways. A grid of individual presence sensors distributed throughout the floor is perhaps the most direct approach. Each of these sensors can be designed to detect pressure on the floor, light, or electrical fields affected by the person's body. Each sensor opens or closes an electronic switch causing the array of switch values to be read into the computer. A person appears as a number of ones in a field of zeroes (Fig. 6.1). This approach makes the computer sensitive to the

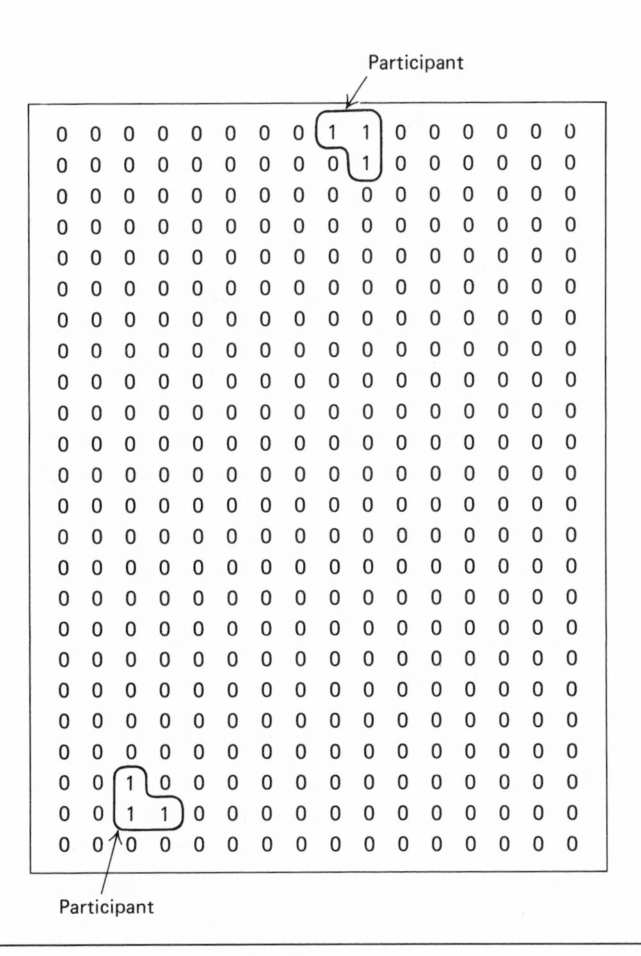

Fig. 6.1 Participants' feet are seen by the computer as ones in a field of zeroes

location of the participant's feet in the room, while completely ignoring any other aspect of behavior, such as arm waving. Velocity and direction of movement, however, can be inferred from a succession of grid samples. If a number of people are present at the same time, they can be discriminated only as long as they are not standing too close together.

An alternate method of determining position is triangulation, computing a person's location based on distance from each of three points. Distance from a given point can be determined in several ways. First, if a light, radio, or ultrasonic source were worn the amplitude of that signal at the sensor would be a function of distance from it. By giving each person an emitter with a distinguishing frequency of light or sound, such a system could keep track of several people at a time. However, there could be a problem if one person inadvertently blocked another's signal. This could be avoided if the sensors were mounted near the ceiling.

A second method of determining the distance from a given point is to emit a pulse of ultrasonic sound and have it repeated by another transmitter worn by the participant. The time delay from the initial pulse to the arrival of the echo back at the emitter is a measure of the distance from the starting point.

Another inexpensive, easy-to-install system operates by the occlusion of light. Distributed around the room are omnidirectional light sources and detectors (Fig. 6.2). The light sources are turned on one at a time and the sensors are read. Wherever light fails to reach a sensor, it can be assumed to have been blocked by one or more participants who are positioned somewhere along the line joining emitter and sensor. To determine the exact position along this line, another light is turned on and presumably other sensors blocked. This second set of readings should pinpoint the person's location. However, if several people are present, the scanning procedure and the computer analysis become more tortured. This system breaks down if more than two or three individuals are involved in the experience.

Motion Detection

While movement can be inferred with position sensors comparing positions at different points in time, motion can also be sensed directly. Ultrasonic sound can be projected into the Environment, bounced off the participant and detected by the computer. If the person is moving, there will be a doppler shift in the frequency of the received signal. The frequency of the doppler signal varies with the velocity of the person's movement and its amplitude varies with the size

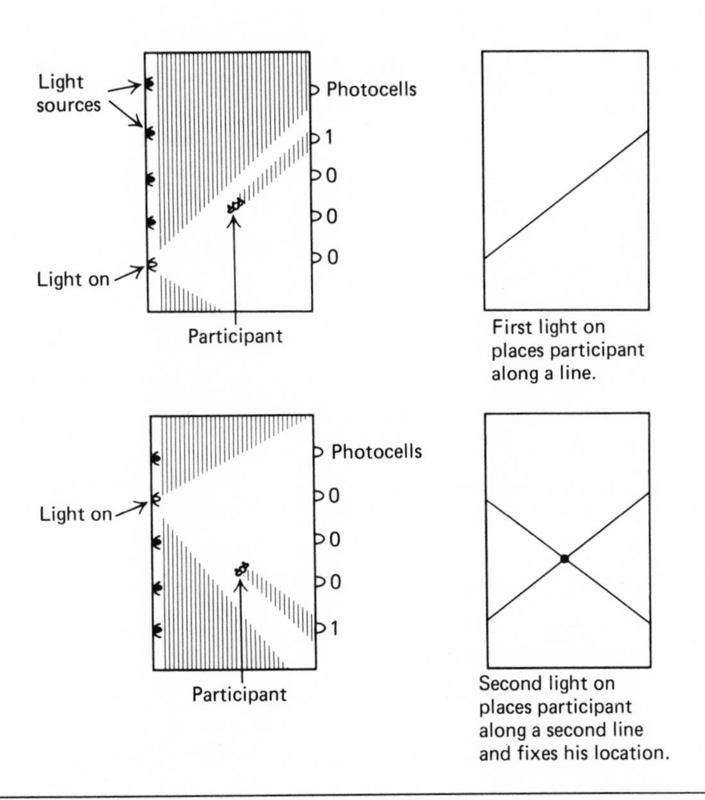

Fig. 6.2 Light occlusion sensing

of the body part being moved and its distance from the sensor. Multiple emitter sensor pairs can give the velocity component in each of several directions. A rough imaging capability would even be possible if special ultrasonic transducer arrays were fabricated. These would be similar to the arrays currently being developed for medical diagnosis.[3] However, such systems may become confused if more than one person is present. Also, spurious ultrasonic signals can be generated by the jingle of keys and coins in a participant's pockets.

Video Sensing

While all of the above approaches present interesting possibilities, a video sensory system is probably most promising. Theoretically, a tremendous amount of information such as the participant's posture, including the position of the head, angle of the torso, and disposition of the limbs can be extracted from a video image. Information about position in the room, height, direction faced, and facial expression is also available. The problem is how to convert this information into a format the computer can perceive and interpret. While there is a well-established field of research in computer vision, the problems are very difficult. It would seem that the Responsive Environment's requirement for analysis of moving images in real time demands too much.

Fortunately, a Responsive Environment is a reasonably benign place to attack the problems of visual pattern recognition. The participant is the only object of interest in an otherwise barren field of view. It is easy to find the participant because the blank walls appear black on the screen and the participant appears as the only bright part of the image.

A simple circuit can take the video signal which consists of many levels of grey and turn it into a high contrast image where every point is black, indicating the background, or white, indicating the participant (Fig. 6.3). It is also easy to detect the edges of the person's image to digitize their outline (Fig. 6.4). Note that much of the important information about a person's outline, and therefore posture, is contained in the first and last edge of the image on each line (Fig. 6.5). The advantage of the abbreviated outline is that there is much less information for the computer to analyze and, equally importantly, it can be stored in a regular format and processed by simple code.

To achieve totally accurate analysis of the person's image, it will ultimately be necessary to process the complete outline and the full grey scale image as well. The amount of processing required exceeds what any single processor can do in real time. Therefore, a number of specialized processors will operate in parallel. Each will be tailored to a particular class of questions. Some will analyze the outline extremes, looking for the head, hands, and feet. Others will do grey scale processing looking for face, hair, hands, clothing patterns, etc. Others will focus on dynamic information. Has the

Fig. 6.3 Grey-scale and high-contrast images

participant moved? How fast? Which part of the body moved? How far is the participant from a particular object on the screen?

Two such systems operated as a stereo pair could be used to infer depth information by comparing the coordinates of features in the two images. While systems such as these provide additional information about the image, they imply a still greater processing load, for the determinations they facilitate require considerable interpretation before the act of perception is complete. More will be said later about the software aspects of sensing.

VIDEO SENSORS: ADVANTAGES AND DISADVANTAGES

These video sensing systems have the advantages of versatility and of portability. They also have some ability to focus

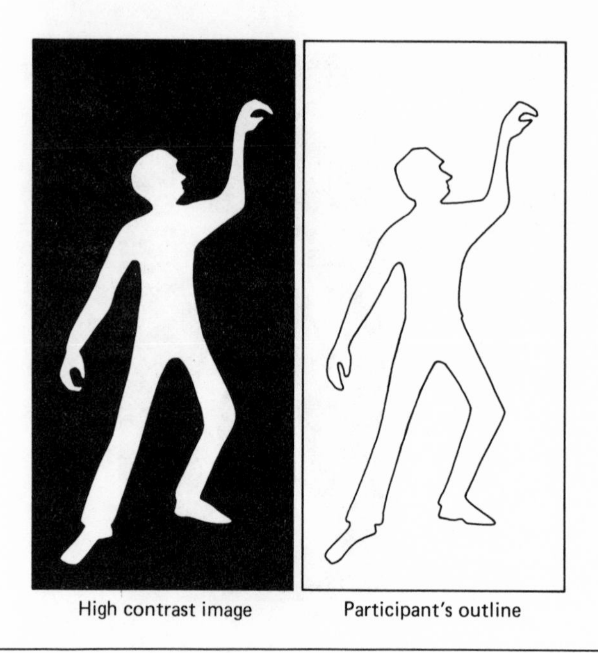

Fig. 6.4 High-contrast image and participant's outline

on parts of the scene. By connecting the camera with motors for zooming, focusing, and aiming, systems could be used as gross motion detectors, or focused on a small area to allow interpretations of subtle movements of the hand or face. Video sensing is pleasing aesthetically. The participant can manipulate the Environment by handwaving, the traditional gesture of sorcerers, magicians, politicians, and college professors.

There are also disadvantages. Because of the effects of perspective, the size of a participant's image changes as he or she moves closer or further from the camera. More problems arise when several people are present as the image of one person can partially occlude that of another, making it extremely difficult for the computer to resolve them as separate figures (Fig. 6.6). This can be alleviated by two means. First, the computer can notice the clothing the participants

Fig. 6.5 Video outline sensor

are wearing and use this knowledge to discriminate between them. Second, additional cameras can provide depth information that may be sufficient to tell them apart. However, since both of these expedients are difficult to accomplish in real time, video sensing is most successful when there is only one participant in the camera's field of view.

MISCELLANEOUS SENSING MODES

The sensory systems described thus far are general systems that could be used to promote many different interactions. The perceptual systems that follow are more limited in scope and suggest more specific interactions.

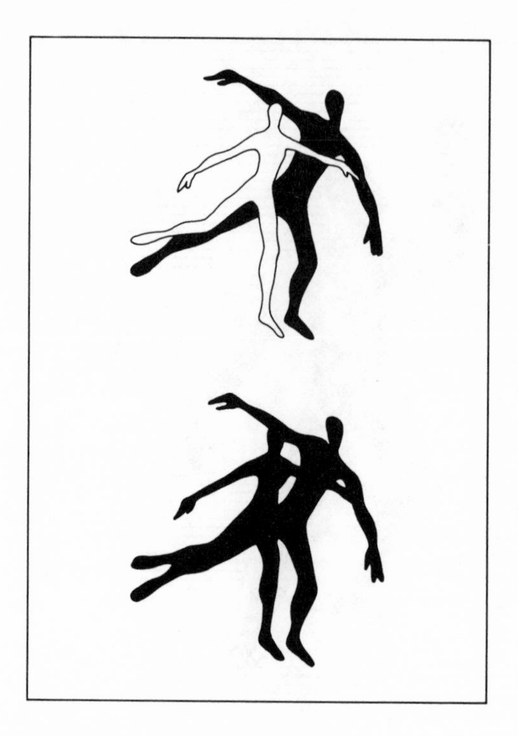

Fig. 6.6 Computer fails to resolve occluded figure

Touch

An alternative to the previous methods for sensing the position of the participant in the room would be a perceptual system sensitive to the participant's touch. Touch sensors would be placed on the walls, ceiling, or on objects within the space. These sensors would be passive, waiting for the participant to activate them. It would be desirable to have sensors that could indicate the amount of pressure as well as the fact of touch so the system could distinguish among taps, touches, caresses, and blows.

Tilt and Direction

Sensors that detect the participant's posture and the direction he or she is facing present additional possibilities. For example, the angle of the head could be determined by equipping a participant with a cap containing a chamber partially filled with mercury (Fig. 6.7). The interior surface of this chamber would contain a large number of electrical contacts. At any time, some of these would be shorted together by the mercury, which conducts electricity. If the head tilted, the mercury would touch new contacts indicating the new direction and degree of tilt. This information would then be transmitted to the computer. The tilt would be extremely useful in an interaction playing with the participant's sense

Fig. 6.7 Mercury cap

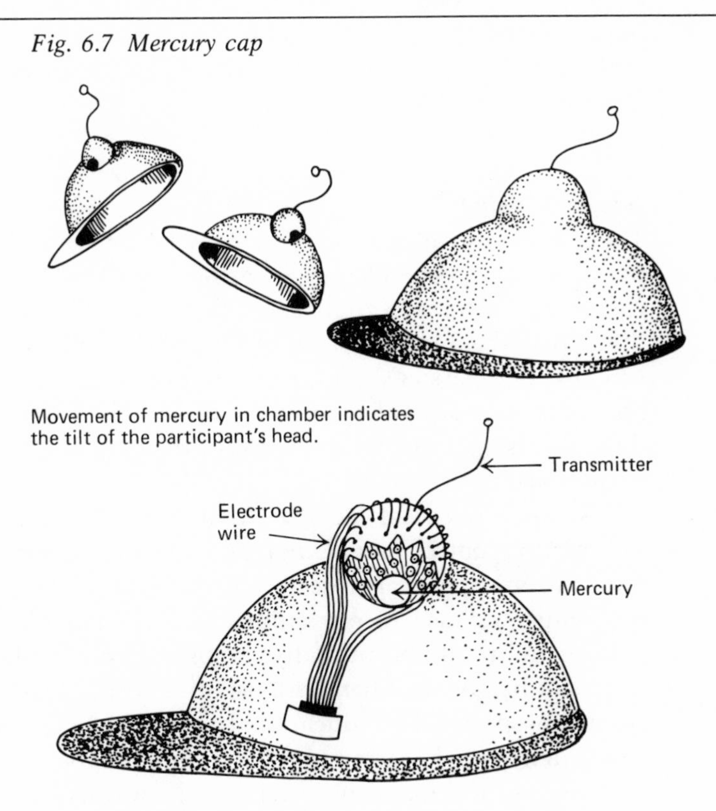

of balance. If the displays defining the visual space were altered so that visual cues conflicted with the perceptions reported by the tilt sensor, the result would be disorientation. If the cap were the only sensor, participants' experience would become quite unusual with their attention increasingly focused on the movements of the head.

Knowing which direction the participants were facing would enable the computer to generate visual displays that would only be visible through their peripheral vision. Thus, as participants turned to look at the display, it would move away. It would remain just out of view, taunting, visible only through the corner of their eyes. A directional light or radio emitter worn on the participant's head would allow sensors around the room to provide this information. A more elaborate system could monitor eye movements by reflecting a beam of invisible infrared light off the eyeball. The source, sensor, and a transmitter could be quite small and easily mounted on special eyeglasses.

Sound Sensing

There are several aspects of a person's voice that can be perceived and responded to, creating more possibilities. The easiest to detect are pitch and volume. Only slightly more difficult is the analysis of the ancillary frequencies that distinguish one person's voice from another's. The recognition and understanding of speech, on the other hand, is a very difficult research problem, except in the case of very limited vocabularies.

As mentioned earlier, the participant's voice is used as a complex control signal, not as a means of verbal communication. Speech per se is not encouraged in the Environments, because in our culture speech and physical involvement are, if not mutually exclusive, usually alternative modes of behavior. When both occur simultaneously, the focus is on speech. For educational applications of the Environments, speech understanding might be desirable. The decision not to discuss it here is partly aesthetic and partly one of priorities. Were such a real-time capability available at low cost, I would put my reservations aside and see how it could be used.

Involuntary Body Functions

Body signals such as skin resistance, heart beat, brain waves, and muscle potentials can be picked up by electrodes attached to an individual. These signals would be quite difficult for the participant to control while in motion, particularly because extraneous signals may be generated as artifacts of physical movement. However, for someone with advanced yoga or meditation training, physically interacting with an external Environment defined by their own internal environment would be interesting. Most people would have to be seated before they could concentrate sufficiently to learn how to control these signals. In this case, the interaction might allow them to perceive apparent movement in the displayed Environment in response to their physiological control rather than their physical movement.

SOFTWARE ISSUES

The data provided by any hardware system will require further processing by the computer before the act of sensing is completed. Among these computational requirements are the needs to insulate the program from faulty sensors, the translation of the inputs into the format of the interaction, the need to keep track of several individuals and the need to detect events that are important to the interaction. Most of the examples given will be drawn from experience with the floor system used in PSYCHIC SPACE. However, any of the other sensing systems described above would have its own computational quirks and special requirements.

Reliability

In PSYCHIC SPACE the first step in the input processing insulated the program from the (occasionally manifest) possibility that one of the floor switches would become stuck, falsely indicating the presence of a person. To combat this kind of malfunction, the program responded not to the state of the floor but to changes of state. When the floor array was read in, the computer compared it to the previous reading to yield two new arrays: the first containing only switches that had just turned off, and the other, those that had just turned on. These two arrays were then used selectively to set and clear corresponding bits in yet another array which hopefully mirrored the true state of the floor minus stuck switches. Every few minutes, this composite array was reset to zeroes, rendering it again impervious to newly stuck switches. Along the way, this procedure isolated the events that were to be of importance to the interaction in the two arrays: one for new switches on, another for old switches off. This information allowed the computer to respond to either footsteps or "unfootsteps" (Fig. 6.8).

More Than One Person

Given the presence of a number of people, the sensory program is called upon to separate the input signals to represent distinct individuals and to assign the most likely identity to each based on prior location. Currently, the programs base their decisions on the distances separating these inputs from the previous locations of identified individuals. The distances can be recorded and responded to if desired.

If the separating distance is too small to isolate the individuals with certainty, the sensing program can report the ambiguity as an aesthetic event which the controlling program can recognize with a distinctive response.

There are several other situations where the computer will be confused. With a floor sensor, a participant can disappear by jumping in the air or climbing on another person. With a video sensor, the participant can hide behind someone else. Alternatively, one person may fool the sensing floor by

Fig. 6.8 Floor sensing

```
0 0 1 1 0 0 0 0
0 0 0 1 0 0 0 0
0 0 0 0 0 0 0 0
0 0 0 0 0 0 0 0
0 0 0 0 0 0 0 0
0 0 0 0 0 0 0 0
0 0 1 1 0 0 0 0
0 0 0 1 0 0 0 0
0 0 0 0 0 0 0 0
0 0 0 0 0 0 0 1
```

Floor as it is

```
0 0 0 0 0 0 0 0
0 0 0 0 0 0 0 0
0 0 0 0 0 0 0 0
0 0 0 0 0 0 0 0
0 0 0 0 0 0 0 0
0 0 0 0 0 0 0 0
0 0 1 0 0 0 0 0
0 0 0 0 0 0 0 0
0 0 0 0 0 0 0 0
0 0 0 0 0 0 0 0
```

New switches on

```
0 0 0 0 0 0 0 0
0 0 0 0 0 0 0 0
0 0 0 0 0 0 0 0
0 0 0 0 0 0 0 0
0 0 0 0 0 0 0 0
0 0 0 0 0 0 0 0
0 0 0 0 0 0 0 0
1 0 0 0 0 0 0 0
0 0 0 0 0 0 0 0
0 0 0 0 0 0 0 0
```

Old switches off

```
0 0 1 1 0 0 0 0
0 0 0 1 0 0 0 0
0 0 0 0 0 0 0 0
0 0 0 0 0 0 0 0
0 0 0 0 0 0 0 0
0 0 0 0 0 0 0 0
0 0 1 1 0 0 0 0
0 0 0 1 0 0 0 0
0 0 0 0 0 0 0 0
0 0 0 0 0 0 0 0
```

Virtual floor minus
stuck switches

lying down and closing switches far enough apart for the resolving program to conclude that there are two people.

Software Sensing

As the systems get more sophisticated so do the demands on the sensing system. It is not enough simply to detect the participant; it becomes necessary for the computer to understand the participant's behavior as a human would. For example, if a person stood facing away from the camera, with an arm extended to the side, and then moved it in front of his or her body, a human viewer would understand what had happened to the arm even when it was no longer visible (Fig. 6.9). A sensory system, on the other hand, might not. It would simply sense the presence of an extended object and upon its disappearance report it had ceased to exist. Similarly, if the outline extremes are being used and the participant raises both hands, movements of his or her head are invisible to the computer (Fig. 6.10).

In addition, the video outline sensor has to deal with visual ambiguity if it is to understand what a person is doing. In the outline sensor's limited vision, a given outline may be consistent with several different postures and in most cases it cannot tell if a person is facing towards or away from the camera.

If the person bends over, lies on the floor, or does a somersault, the outline may provide no clue about what is going on. In these cases, more sophisticated processing must be applied. It may be necessary to acquire more detailed information than the outline extremes. Grey scale information may be required. However, analysis of the outline can serve to focus the grey scale processing so it is only applied to a small part of the image. Once the computer has recognized the head from the outline, it can apply special processing to the area inside the head to see if the participant is facing the camera. If it cannot identify the head from the outline, special texture operators tuned for identifying hair may be applied inside the outline.

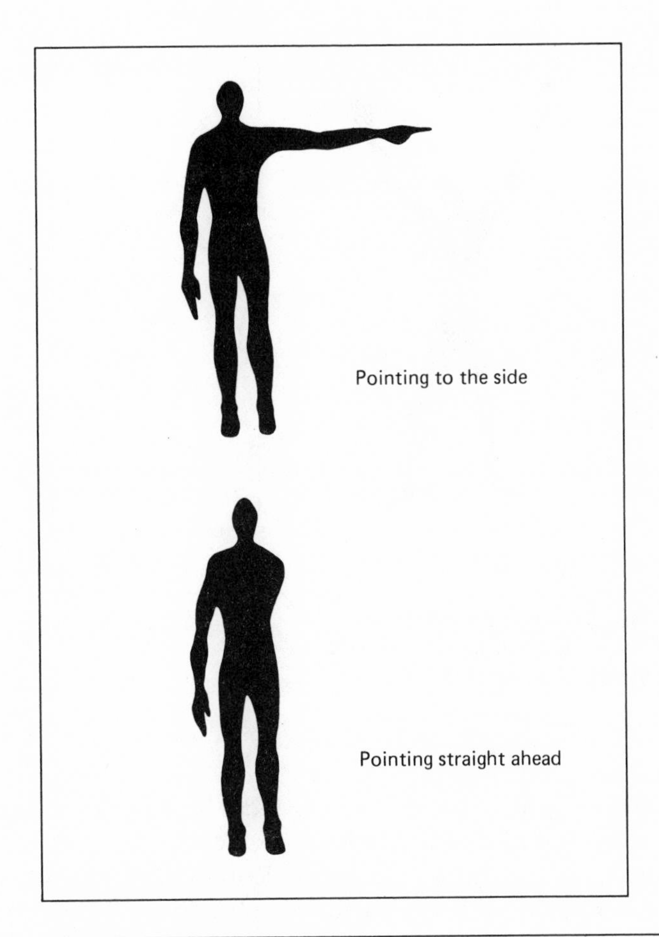

Fig. 6.9 A problem in software understanding of human movement: what has happened to the arm?

In the event that an isolated outline is ambiguous, the sequence of outlines leading up to it often resolves the confusion. In general, it is a good idea for the software to take special notice when an unambiguous posture occurs. With this high confidence point as an anchor, it can predict likely movements and eliminate from consideration interpretations that are consistent with the input, but not physically pos-

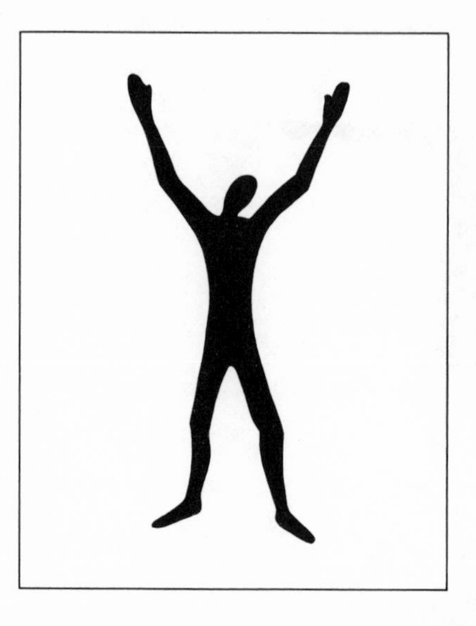

Fig. 6.10 Movements of the participant's head are invisible to the computer

sible.[4] To do this, the system must be armed with commonsense knowledge about how people can move. A number of systems for modelling the articulation of the human body exist and could perhaps be adapted to use in real-time pattern recognition.[5]

Understanding the person's image frame by frame is not enough. It is actions that are important. The system should be able to distinguish between casual movements and meaningful gestures such as an affirmative nod. It should be able to distinguish between walking and jumping. To make such discriminations, it must ultimately be able to see in the fullest sense. Until this is possible, the system must make the most intelligent use of the input and processing power available to it.

Once the behavior of the participant is understood in isolation, it must be considered with respect to the context that the computer is defining. An action that has no significance in one context may represent a dramatic event in another.

In more sophisticated interactions the computer must interpret the participant's behavior in terms of an internal model of the Environment that it is displaying. In PSYCHIC SPACE, the computer viewed a person's movements on the floor as the movements of a symbol through the Maze. Each step had to be tested against the model of the Maze to make sure the participant's symbol had not crossed one of its boundaries. In more advanced systems, there will be interaction with a completely displayed world varying with the participant's apparent movement within it. In this case, the computer must sense the relationship of the participant's image to the displayed Environment.

One final but critical observation about these perceptual systems must be made. They will all fail. Each can be fooled in certain cases. However, their goal is not necessarily to report events perfectly as they occur. Rather the intent of these systems is to represent the act of perception itself and to suggest by their strengths and weaknesses just what kind of creature the Environment is and therefore what kinds of actions are of interest to the participants, who will be exploring the computer's perceptions just as surely as they are exploring the displayed Environment.

NOTES

1. "Focus on Light" (Trenton: New Jersey State Museum Cultural Center, May 20–September 10, 1967).

2. B. Raphael, *The Thinking Computer* (W. H. Freeman, 1976). pp. 275–86.

3. "Sophisticated Medical Ultrasound Imaging Made Possible by High-Speed Data Converters," *Digital Signal Processor* 1, No. 5 (August 1980) (TRW LSI Products).

4. F. Hayes-Roth & V. Lesser, "Focus of Attention in the HEARSAY II Speech Understanding System," *International Joint Conference on Artificial Intelligence* (1977):27–35.

5. Norman I. Badler & Stephen W. Smoliar, "Digital Representations of Human Movement," *Computing Surveys* 11, No. 1 (March 1979):19–38.

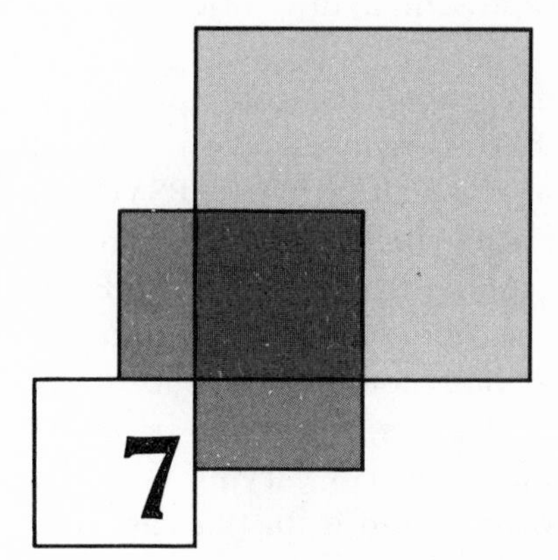

THE ENVIRONMENT RESPONDS

THE DISPLAYS

Responsive displays have a variety of functions in an Environment. The responses establish relationships between participants' actions and what they perceive as well as forming expectations about the consequences of future actions. Each successive response either verifies the current relationship (between the person and the Environment) or signals that the rules have changed.

The displays also define the context of the interactions. The Maze in PSYCHIC SPACE is an example of this. In the distant future, computers may be able to create such effective multisensory domains that a participant will feel that the displayed space is reality. Artificial creatures that interact with participants or with their representation on the display screen will be introduced to strengthen this illusion. Such intelligent, behaving entities will be cousins of the familiar mechanical robot.

AESTHETIC ISSUES

Display feedback is used to encourage general activity and to motivate specifically desired behavior. In an Environment where nothing happens unless the participant moves, each response is a reinforcer that promotes further activity. Lower levels of activity can be induced by having the Environment cease to respond, delay response, or respond only after progressively longer delays. Just as a painter has techniques that guide the observer's eye to important parts of a painting, the Environmental artist will sometimes wish to provoke very specific actions, such as having the participant move to the starting point of the Maze in PSYCHIC SPACE. In that dark Environment the desired action was suggested by displaying an unexplained object on the screen motivating the participants to move their symbol to the starting point. Sound can also be used to guide the participants, as in PSYCHIC SPACE when the Environment only responded to movements that led them to the phosphorescent panels.

Responses can also be used to define an identity for each participant. In the Maze, a symbol represented one participant. Two people could be represented by different symbols. Or, one person's footsteps could provoke responses in sound, while another's footsteps affected patterns of light. The only essential criterion is that distinguishing feedback follows each participant's actions.

Computer Compatibility and Scale

There are a number of requirements that must be satisfied by any display. At the very least, it must be amenable to computer control and designed to take advantage of the computer's capabilities. It must respond at computer speeds and have a large vocabulary of possible responses. The tubes in GLOWFLOW failed by both of these criteria because the phosphors were slow to respond and the number of possible patterns severely limited by the arrangement of the tubes. It is economical for a display to be general enough to express a large number of relationships around different themes: it then becomes a tool for exploring a whole domain rather than a single piece.

It is also advantageous for the display to use the whole space as this encourages the participant to move freely. Visual displays should be large enough to give the feeling that they are the space, rather than objects within it. The tubes in GLOWFLOW were successful by this criterion, while the video projections of METAPLAY and PSYCHIC SPACE were less so, as the images and therefore the interaction were focussed toward one end of the room. Although the projection screen was large enough to become the end of the room, it would have been preferable to have displays on all the walls as well as the floor and ceiling.

The sound system should be capable of placing the perceived origin of sound anywhere in the space, or at a specific point above, below, or beyond its walls. It should be possible to create multiple sources at different locations. The purpose would be to create a precise sense of auditory space. This goal mandates a large number of speakers as well as special processing to pinpoint the origin of the sound.

Since the conditions for display are as important as the display itself, it may be desirable to control the space completely so that the display is the only source of stimulation. In this way, visual displays are contrasted with darkness and sounds with silence. Against this background of sensory deprivation, the displays not only appear more dramatic, but their stimulation becomes necessary for the participant's psychological comfort.

Complexity

The issue of complexity should be considered carefully when programming displays for this medium. Clearly, the displays and the relationships they define must be complex enough to be interesting. However, it is important to note that the participant must continually:

1. Process the Environment's most recent response.

2. Interpret the relationship indicated by that response.

3. Choose the next action.
4. Predict its consequences.
5. Effect that action.

Much of the participant's mental capacity is consumed by this processing. Therefore, the light and sound information need not be as complex as they are in passive art forms where the audience just watches or listens.

In fact, there is a definite danger of sensory overload leaving the person unable to understand all that is happening. In other media and especially in what is called multimedia, the idea of overload is in vogue. This trend may reflect a boredom with conventional media and frustration with their inability to surprise. However, within the Environments it is simple to surprise. The problem is to provide enough structure to define the relationships that, when broken, create information.

We have to introduce the whole idea of composed response and to define relations patiently; only when they are apprehended can we begin to play with the framework of the medium. In a mature medium it would be necessary to break the rules of informed expectation in order to communicate. On the other hand, the rules must be there to be broken or the possibility for information does not exist. Our first task then is to define the initial rules.

Realism Versus Imagination

One tendency, which is not necessarily desirable, is the use of displays to reproduce the real world. The technical evolution of television suggests that other displays will undoubtedly progress toward greater and greater realism. But it is probably an error to try to automatically guide aesthetic displays in this direction. An abstract conceptual or symbolic space may be more effective than a completely faithful rendering of a real environment. What is important is that the displayed space appear sufficiently compelling so that the participant suspends belief and accepts the experience as real, even if the world it portrays is not.

There are, of course, more practical reasons postponing representationalism. Computers currently are incapable of generating fully realistic displays in real time. Line drawings and simplified representations of solid objects are the best that can be expected, although real-time three-dimensional video graphics will improve dramatically in the near future.

VISUAL EXPECTATIONS

The most important and best understood sense is vision, which more than any other provides us with our sense of place. When visual data conflicts with that of other senses, it usually dominates. This dominance is partly the result of very simple expectations based on a lifetime of experience. The following example is taken from personal experience and is offered to indicate the aesthetic importance of these conventions.

Mystery Ride

A number of years ago on the Boardwalk at Ocean City, New Jersey, there was a concession called the Mystery Ride. A group of about ten people entered a very ordinary-seeming room complete with pictures on the walls, a light fixture hanging from the ceiling, and a rug on the floor. We were seated on two benches suspended by heavy beams from either end of the room. When the ride began, it seemed we were being turned upside down. This was alarming as there were no physical restraints to keep us on our benches. After a panicky moment, I reasoned that it was the room that was turning around us, while we remained stationary. Unfortunately, this intelligence was not at all reassuring, for years of experience tell us that ceilings are up and floors are down, and when one finds his head by the floor and his feet by the ceiling, he assumes that he is upside down (Fig. 7.1).

I closed my eyes to rely on my vestibular sense. The semicircular canals, which had served only to make me seasick in the past, were now being asked to offer evidence in

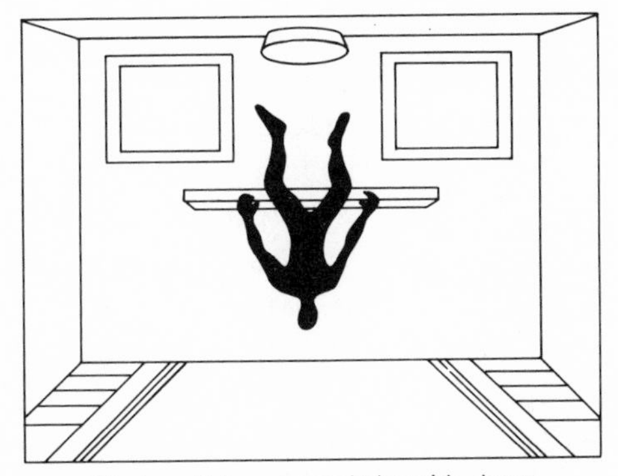

Illusion: Being suspended upside down

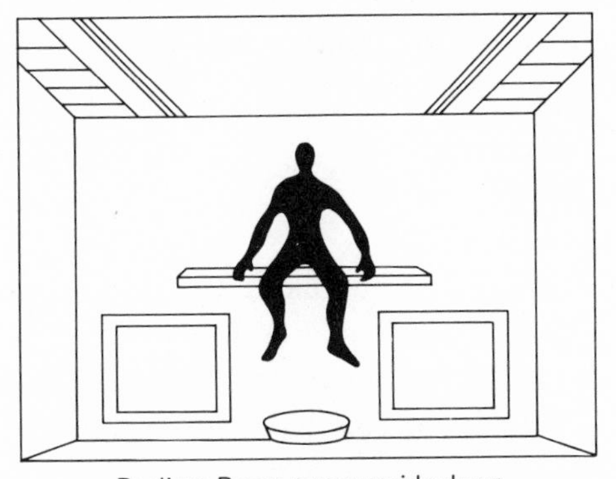

Reality: Room turns upside down

Fig. 7.1 Mystery ride

support of intellect in its case against untrustworthy vision. Unfortunately, the authority of the visual interpretation made the interior sense unreadable. While I was familiar with the psychological literature describing visual illusions, I had never guessed how dramatically vision was able to tyrannize the other senses and overrule reason as well.

This illusion suggests that there are a number of conventions established by our visual experience that can be exploited for use in an Environmental display. More than the other senses, vision defines reality or unreality. We can become disoriented and even unbalanced if the relationships anticipated by our sense of vision are altered.

Continuity

Some of these visual expectations are the result of the consistent experience that accompanies movement in any physical space. As we turn our head or move around, we are accustomed to seeing one scene transformed into the next in a very smooth and connected way. Visual feedback could be made to deviate from this expectation in a number of ways. Participants' movements in the Environment might result in their appearing to move in the visual space in a disconnected way. Or, the rate of movement could be exaggerated so that a three-foot stride resulted in an apparent thirty-foot change in position on the display. Thus, by physically moving around the small space of the Environment, a person could explore a much larger one perceptually.

Perspective

In addition to continuous feedback, we expect that when we approach an object it will become larger and when we move away from it, it will become smaller. This relationship could be reversed, so that as one moved toward a displayed scene it would appear to recede, whereas, if one reversed direction, it would become larger. This suggests a whole family of translations of the physical space into a perceptual space. By moving in one direction people might cause the perceptive Environment to rotate around them rather than to approach and recede. Not only could an individual seem to move about a much different sort of space, but the space could also be perceived to move around the individual.

Vision and Memory

Since our sense of vision expects continuity, it relies less on memory than the other senses. As we move around the physical environment, we don't have to remember what we saw a moment before; it is either still there or something derived from it remains. Sounds, on the other hand, are heard and then cease to exist; they must be remembered as they are interpreted. Perhaps for this reason there is no tradition of composing sequences of abstract visual sensation as there is for sound, although John Whitney's computer-generated films of abstract images suggest that such an art form may be possible and effective.[1]

MEDIA

Discrete Lights

A variety of media could be used to construct the displays. Arrays of lights are one possibility. The lights activating the tubes in GLOWFLOW were under independent computer control. The distortion of space that they created could be amplified by using a large number of linear light sources. Extremely narrow neon tubes, electroluminescent strips or optical fibers could define a number of alternative visual spaces (Fig. 7.2). Each subset of the visual lines would define

Fig. 7.2 Space dance

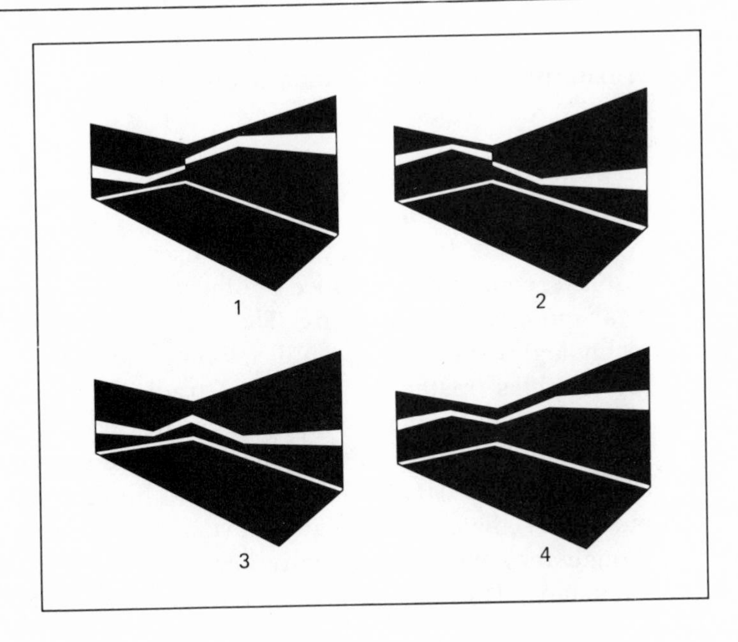

a particular illusory space within the darkened Environment. By switching different groups of lights on or off, the space itself would be articulated, seeming to move around the participant in a SPACE DANCE. While such an Environment would be very effective, it would be limited compositionally to the single theme of spatial distortion.

An alternative use of discrete lights would implant a 64 × 96 array of discrete lights such as LED's (Light Emitting Diodes) into the floor. Such a display would be extremely powerful because it would bring the responses close to the participant, making an interaction like the Maze more immediate. The tremendous number of lights makes a rough approximation of graphic capabilities possible. For example, the participant's trail of footprints could light up, graphic creatures could be chased around the room or graphic waves could flow along the floor and break around the participant's feet.

Disco floors are an example of an effective use of discrete lights. However, while they are dramatic, these displays lack the resolution to create anything but rough patterns. In addition, they receive only minor attention from the dancers as they do not respond to motion and must compete with many other sources of stimulation. They are part of an overload approach, rather than a disciplined and structured one.

Marriage of Computer and Video

While discrete light arrays can be used to create effective displays, computer and video graphics have the potential for a far more varied repertoire. There are three major display technologies that have been developing independently: vector graphics, raster graphics, and traditional video image processing. Each of these approaches has been progressing rapidly and promises increasingly dramatic displays in the future. Each has its own strengths and limitations. For the most part these technologies have remained separate although they have occasionally been combined in film and video tape.[2] However there are a number of technical reasons why these different methods are likely to merge. Each will be discussed separately and then a new, computer-compatible, color display medium, based on a marriage of their strong points, will be described.

Vector Graphics

The display technique most commonly associated with computers is vector graphics. In vector graphics, pictures are created by point-to-point plotting of lines on a cathode ray tube. On sophisticated systems, thousands of vectors can be displayed. Gray scale can be controlled to some extent and analog hardware can accomplish the real-time movement, scaling, and rotation of image elements. With these capabilities, linear representations of scenes can be created and manipulated in real time. Although the image is only a two-dimensional projection of a three-dimensional scene, if the displayed scene is altered appropriately as the viewer moves, the sense of depth is very strong. This effect was explored during PSYCHIC SPACE by projecting simple perspective scenes and altering the viewpoint as the participant moved about the room.

An alternative method of handling the vector screen is to draw "key frames which represent a scene at significant points in a sequence, and then to use the computer to generate the intervening frames by interpolation. Smooth animation can be achieved in this way and since there are many ways the interpolation can be done, a whole universe of new animation techniques becomes available (Fig. 7.3). Key frame animation was perfected by Burtnyk and Wein at the National Research Council of Canada and was the method used to make Peter Foldes's impressive movie "Hunger."[3,4]

The advantages of vector graphics are twofold. First, only the endpoints of lines have to be computed. Second, hardware exists that will move, scale, rotate, and duplicate parts of the image in real time. In addition, there is a surrealistic quality to projected vector graphics that is useful aesthetically. It is as if the schematic world of diagrams had been brought to life. Ultimately, however, a medium restricted to

Fig. 7.3 Key frame animation

straight-line segments without width, color, or texture will prove confining and a more expressive medium will be desirable.

Raster Graphics

As memory costs have declined, a second display technology has overtaken traditional vector graphics. These new displays are based on the raster (scan line) format of broadcast television technology. As can be seen just by looking at a television screen, the image is divided into discrete horizontal scan lines. Each of these lines is redrawn thirty times a second. In raster graphics, each point on the screen corresponds to a memory cell that contains the visual information to be displayed there, which means that enormous amounts of memory are required for high resolution displays. When the information in a cell is displayed, it is called a pixel, and can range from a single bit, producing a black-and-white image, to fifteen bits, which allows a palette of 32,768 different colors. The image is produced by reading out the pixel memory into digital-to-analog converters which actually generate the color signals. However, once the computer has written information into the refresh memory, it is no longer involved, nor is its performance degraded by the refresh process.

Raster graphics has the advantage that it can be used to generate color and shading at every point on the screen and therefore can create solid and realistic looking images. Very convincing three-dimensional scenes can be portrayed. Complex objects can be approximated with appropriately shaded faceted surfaces. With additional processing, the faceted surfaces can be smoothed and continuously curved and shaded, providing a realistic portrayal of objects (Fig. 7.4).[5] The images produced by some of these displays are extremely effective and even a little spooky. In particular, one film sequence of a talking head is very disturbing.[6] While one is not deceived into thinking that it is a real person, the fact that a bunch of numbers produces a result that raises such a question guarantees that the future holds the ability to create any visual reality or fantasy that we can conceive. There will be no limit.

Animation will be possible with what appear to be live images. Computer-generated "people" will be capable of impossible feats and subject to magical transformations. Unfortunately, the computation that these effects will require will be enormous and for the moment are unthinkable in real time. However, once something can be done at all, it can then be done faster and faster as new technology is applied, and ultimately this new visual reality will be part of the Responsive Environment.

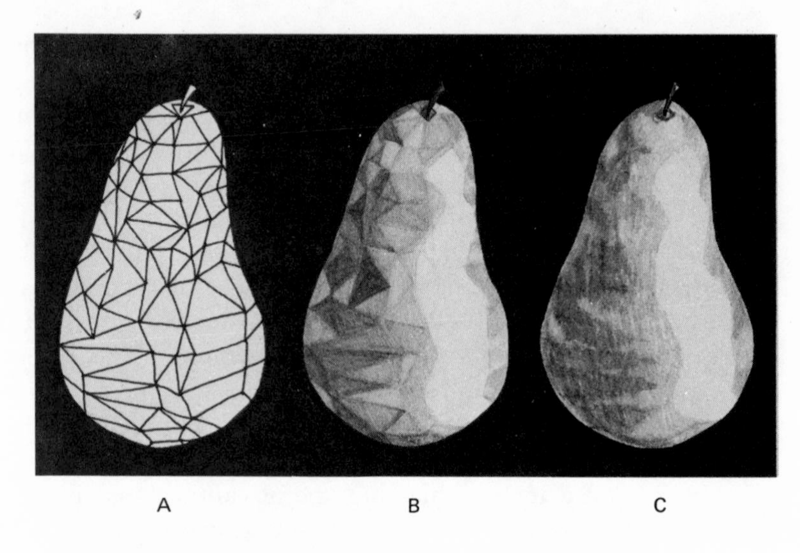

A = Segmented approximation of surface
B = Shaded segmented surface
C = Smoothed shaded surface

Fig. 7.4 Approximations of a surface

Video Processing

As computer display techniques have evolved, there has been an independent but parallel development of television technology itself. Television as a broadcast medium has its own repertoire of techniques, some of which are not available through the other display technologies. Recently, innovation in this area has accelerated so that the repertoire available to the television producer now comes close to providing a composable medium as opposed to a few special effects. Only recently has this medium begun to be influenced by digital technology and computers.

Television technology differs from computer graphics in two very significant ways: first, television can make use of live images, and second, television has always been primarily an analog technology with its own techniques which until recently have been difficult to accomplish with digital circuits.

One of the effects that analog technology makes trivial is crossfading between two images in which both are visible with varying intensities at every point on the screen. A second technique, keying, allows part of an image to come from one camera, while the rest comes from another. For example, if a person stands in front of a neutral background, his image can be separated from that background and inserted into a new scene (Fig. 7.5). Alternatively, the outline of the participant's image can be used to form a window from one image into another (Fig. 7.6).

Fig. 7.5 Standard video keying

Control image
against neutral background

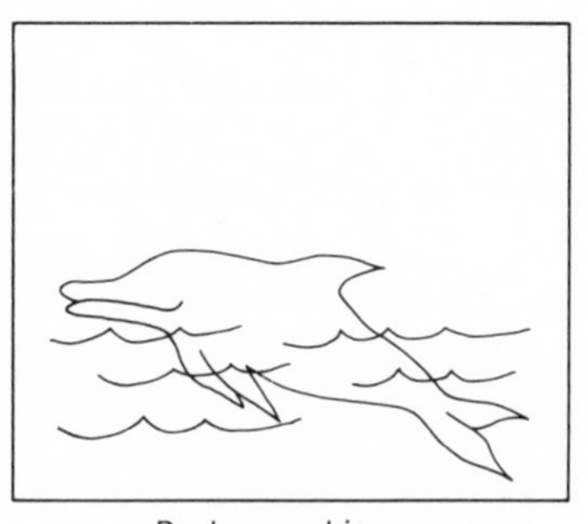

Background image

Composite image

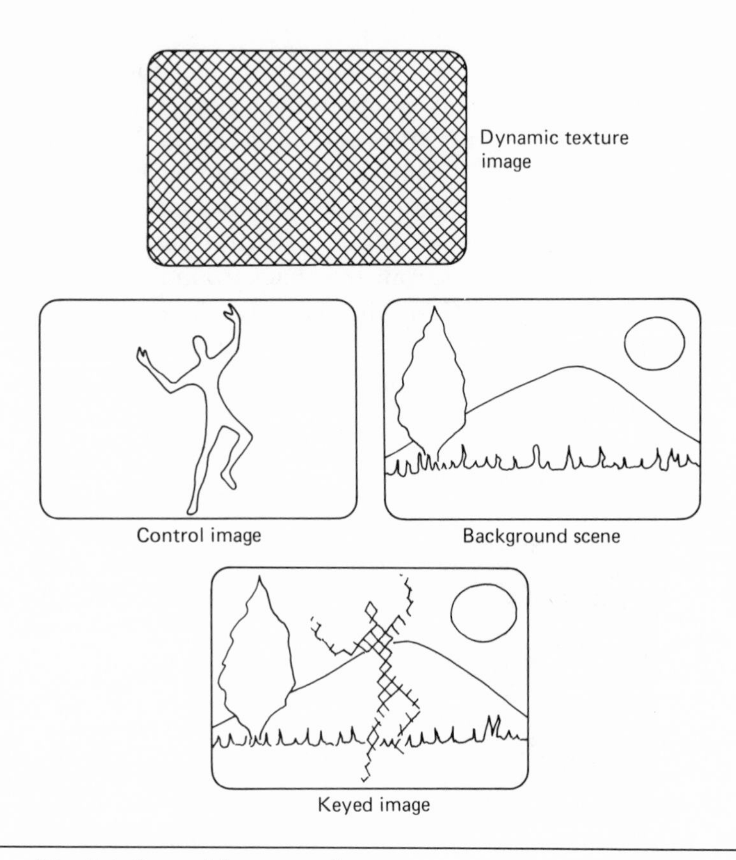

Fig. 7.6 Control image used as a window from one image to another

Video Storage

Another class of video effects depends on the ability to freeze frames and to replay sequences. There are three methods for accomplishing these effects: video tape, video disk, and semiconductor memory. Video tape is the most familiar but the least flexible. It can easily record continuous action and play it back later. By using two recorders, one to record and the other to play back, it is possible to delay a live image by several seconds and project it by itself or superimposed on the undelayed image. This technique poses an interesting problem for viewers who must try to coordinate the actions of their past, present, and future images. This simple effect has been the basis of a number of video art pieces.

Analog Disk

The analog disk is the device used for instant replay in sportscasts. It typically provides storage for up to 500 frames (about sixteen seconds) of continuous video. It can be played back at normal speed, slow motion, step frame, freeze frame, reverse action or loops. Some of these effects are seldom used in broadcast television but can present interesting live interactions. For example, a clumsy action can be made to look graceful if played back a frame at a time. Similarly, an innocuous gesture, such as scratching one's nose, can be made to look absurd when endlessly repeated.

Digital Memory

Another class of techniques are based on the use of digital memory for image storage. A video snapshot can be taken by digitizing and storing a complete frame. Thus, a person's image can be frozen in midair, the video equivalent of GLOW-MOTION. Alternatively, a new snapshot can be taken and displayed every half-second to create the disconnected effect of partial animation.

Since most digital image storage is intended to record the video image faithfully, high resolution is usually desired. However, there is an interesting effect that occurs when very rough digitizing is used. In that case, a person's image is reduced to a series of small blocks which move as he does. The robot's view of the world in the movie "Westworld" was generated in this manner.

This storage technology is rapidly becoming cheaper. It is now possible to consider the storage of sequences of frames. The current system at the University of Connecticut can store up to three high-contrast frames and is being expanded to sixteen. With several digital frames a new category of special effects is possible. For instance, if several images of a person are stored each can be made a different color. Or, a distinctive color can be assigned to each of the areas where the images overlap. This technique can be varied by making the screen dark except where consecutive images do not overlap. If a person stands still, there is complete overlap and the person's image disappears. If the person moves, only the moving parts of the body are visible. Interactions based on just these simple effects are surprisingly engrossing.

This new system will provide much more flexible manipulation of stored images. First, it will not be necessary to store the entire frame. Any area of the screen can be defined as a window and its contents stored, regardless of its shape. Second, stored windows can be moved around the screen as they are displayed. Third, windows can be used to store graphics as well as live images. Fourth, many windows can be displayed on the screen simultaneously. Fifth, image memories can be grouped to provide more grey levels or colors. Sixth, both graphic and live images can be colorized.

A host of visual effects will be possible with this system. Many views of graphic creatures can be stored in different parts of the same memory. Convincing animation can be achieved by sequencing through them as they are moved about the screen. This is how the characters in video games are implemented except that in this case hundreds of high resolution poses for each character will be stored.

In addition, parts of an image can be stored in different windows and broken apart. Many snapshots of a person's face can be saved. In these, faces can fall off the body, or a face remaining on the body can retain the expression the participant had a moment before. Alternatively, other people's faces may be placed on the participant's body. Or, the features from various faces may be combined to create a composite face. Finally, a mouth might be replaced by an eye and a mouth substituted for each eye. Thus the eyes would grin and the mouth blink. These visual effects suggest the development of a video kaleidoscope or an intelligent fun-house mirror.

Video Image Processing

There are also a set of image processing techniques that were pioneered by video artist Nam June Paik and perfected by Dan Sandin who is on the art faculty at the Chicago Circle Campus of the University of Illinois. Dan Sandin's image processor can take the lines from a vector graphic display and turn them into spectacular colored images. His original system was based on simple analog processing of black and white video images. The analog processing consisted of simple addition, subtraction, multiplication, integration, and differentiation of video signals. These can be turned into color either by having separate processors for red, green, and blue or by colorizing a single signal by defining a number of brightness thresholds in the input image and assigning a different color to each interval between thresholds. Sandin has used his image processor in conjunction with a graphics system developed by Tom DeFanti, also of Chicago Circle, to create a live performance graphics medium.

Digital Techniques

Recently, it has become attractive to process images digitally. With colorizing, for example, the total range of grey levels in a video image is subdivided and each of the subranges assigned a color (Fig. 7.7). With digital techniques, the color assigned can be changed every frame. The result is that a person's image can become a mass of color. These colors can then be made to crawl or dance around the person's body. Graphics generated by the computer can be subjected to the same effect.

Another use of digital techniques transforms the most mundane of video effects. A switcher allows the selection of an image from one camera or another by mechanically pushing switches. In 1973 I devised a computer-controlled switcher that allowed twelve video inputs. This device can switch from one image to another to another at frame rates. It can switch between alternate frames in a rhythmic fashion. By placing a number of cameras around a dancer, the visual

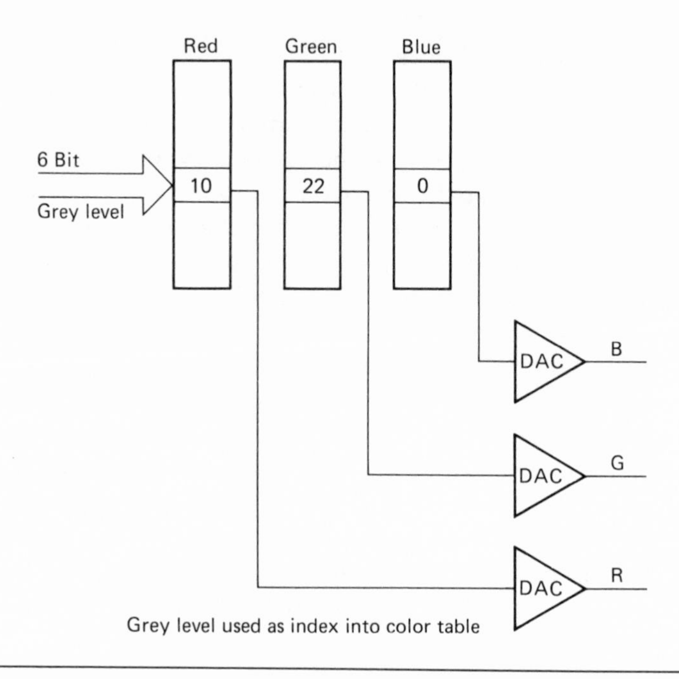

Fig. 7.7 Digital colorizer

perspective appears to dance as well. The effect is extremely arresting and has recently been used by Twyla Tharp, a New York choreographer.

Feedback

Feedback is another aspect of video image processing. It is created by pointing a video camera at a monitor that is displaying the image the camera is picking up. The result is an infinity mirror effect of smaller repetitions of the image within itself. Color video feedback creates far more spectacular patterns than black and white. The complex delays caused by the sophisticated color circuitry make color feedback unstable. Beautiful dynamic imagery can result from small perturbations in the input image. The problem with feedback is the lack of control; pretty images are easy, almost automatic, but to truly conceive and compose is next to impossible with most equipment. Only the scan modulation techniques, to be described, can make feedback and image processing more controllable.

The video medium is developing rapidly. It has the advantage of using the real world as a source of complexity. It also allows the inclusion of participants' images in the Environmental displays. This medium by itself suffers from one surprising problem. The computer can intervene in only the most limited ways because video signals change much faster than the fastest computers can do even trivial computations. Therefore, video effects must be accomplished by hardware and specialized processors. The computer must be content to sit back and orchestrate the behavior of the faster hardware. Most of its decisions will be at frame rates.

COMPUTER VIDEO

Each of the display technologies that have been described has its own idiom and its own unique capabilities, but none is ideal. What is needed is a tool for taking the best features of the existing technologies, adding a few new features and melding these ingredients together in a single medium under computer control.

Vector graphics is fast but limited to schematic scenes. However, by scanning a vector image with a television camera and subjecting it to Sandin's video processing, the lines can be given quality and color. Images from different cameras can be processed differently and merged together in a composite image. Dramatic imagery can be created in this way.

Scan Modulation

It is possible to step outside the confines of the raster to create new video effects. In this case it is the raster itself that is manipulated. The result is a large family of distortions of the video image (Fig. 7.8). Bill Etra and Steve Rutt at New York University developed a video synthesizer based on such

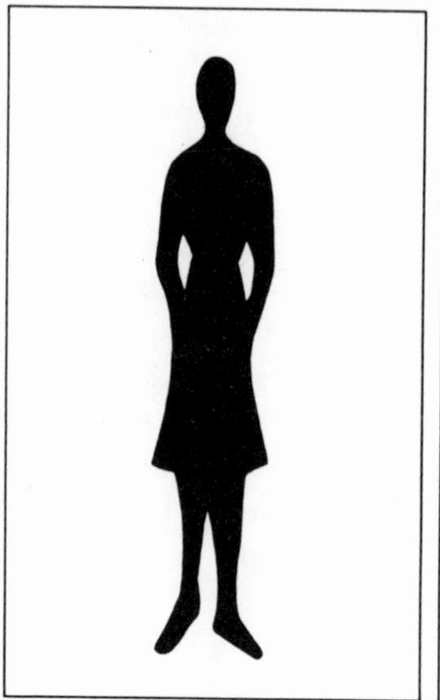
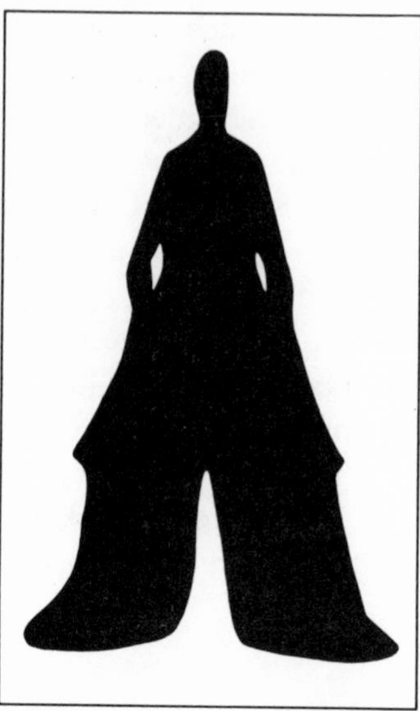

Fig. 7.8 Simple distortion of participant's image by scan modulation

scan modulation. For the purposes of Responsive Environments, a few very structured transformations are far more important than the general ability to bend and distort images in arbitrary ways. The desired manipulations are those required to duplicate the move, scale, and rotation features of traditional computer graphics (Fig. 7.9). With these features available, live images can be treated as image elements with exactly the same status as graphic objects, and entities and scenes can be combined in coordinated ways.

Combining Multiple Elements

Thus far, discussion has been limited to the means of generating separate image elements with no thought about how they interact or whether one is in front of and therefore occludes part of another. The question to be considered now is how to create complex images which are the composite of a number of separately generated image elements.

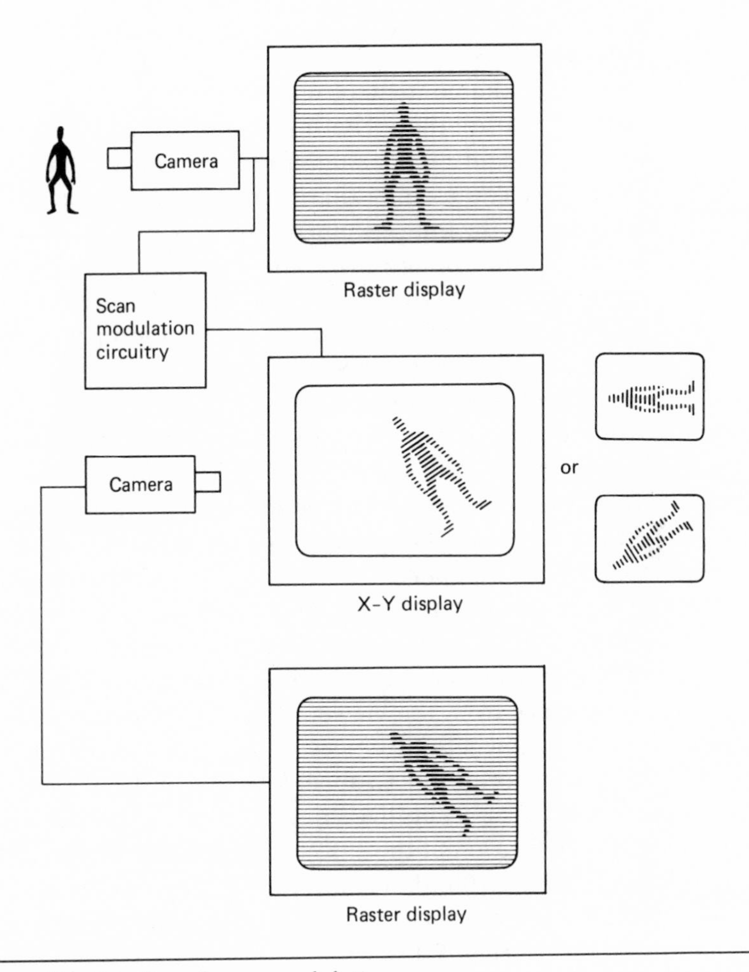

Fig. 7.9 Use of scan modulation to move image element

The reason for concentrating on the separate generation of image elements was to remove the problem of merging elements from the realm of computer processing. These computations are possible in principle, but not in real time. Image elements are kept separate so that they can be generated by separate parallel processors of high speed and specialized design. The combination of the separate image elements can be accomplished easily by a generalization of the keying effects described earlier.

Priority Keyer

In keying, the content of one image is used to control the visibility of another. The high intensity part of one image can be thought of as being of higher priority than the other image. We can assume that there is one place in the system where all of the image elements are brought together in video format. Vector images will have been picked up by cameras, scan-modulated real images rescanned by cameras, and all raster graphics will already be in the appropriate form. All of these video images are now synchronized.

The next step is to decide which of these image elements is to be shown at the points where two or more overlap. In general, we should show the one that is closest to the viewer. This decision must be made at video rates for every point on the screen. Although no standard video equipment exists to perform this particular function, it is not difficult to design. In fact, the idea can be generalized to allow the image shown at each point to depend on the particular combination of elements present at that point rather than a strict ordering by distance from the viewer. Fig. 7.10a illustrates the everyday reality where, if object A is in front of object B and B is in front of object C, then A is also in front of C. On the other hand, a generalized device would allow the independent specification of the case where A and C are present without B (Fig. 7.10b). Very interesting sculptural shapes can be created when several live images, e.g., dancers, overlap using this device (Fig. 7.11).

Fig. 7.10 Priority keying

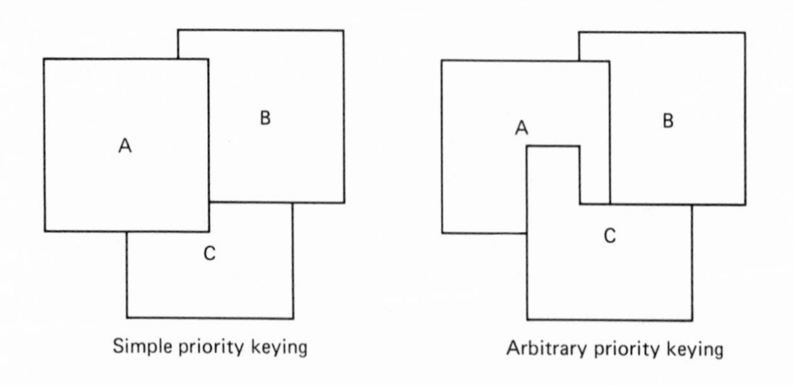

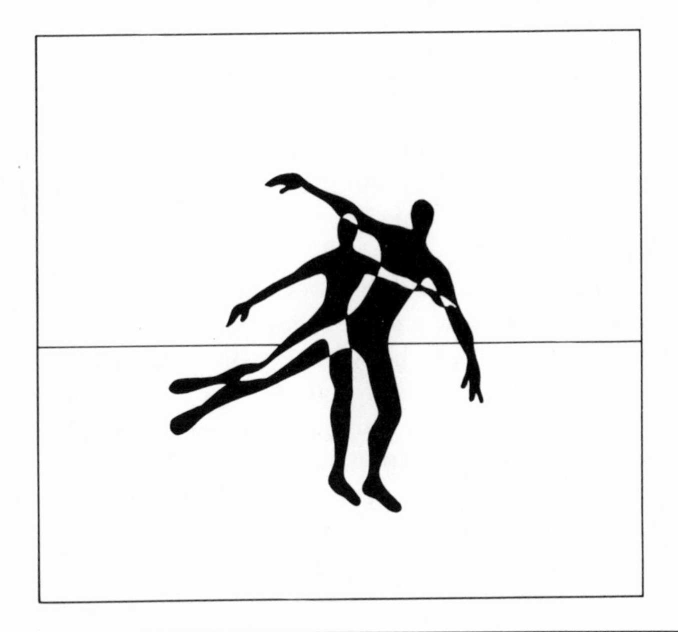

Fig. 7.11 Dancers forming sculptural shapes through priority keyer

The range of processing techniques just described defines a powerful visual medium that is designed to maximize the best of the available technologies and to allow the application of new display generators as they become available.

Image Presentation

Once the image has been generated it must be presented to the participant. In all work so far, a single video projector has been used. It would be desirable to completely surround the Environment with a single continuous image just as reality surrounds each of us. This can be accomplished using a number of projectors as has been done at Disneyland with film of a boat ride.

Another approach would be the one taken by computer graphics pioneer Ivan Sutherland which uses a head-mounted display to provide a separate stereo image to each eye.[7] As a person moves his head, the motions are sensed and the view of the scene displayed updated accordingly. The effect is as if the person were actually looking around in a three-dimensional space. This system does not really allow the viewer much freedom of movement because movements are sensed mechanically, but it clearly suggests a theme that will develop into portable goggles displaying transmitted video signals.

An alternative design would be to fabricate a display on a contact lens and a sensor that would detect eye movements as well as head and body movements. This display must then generate only the image that the eye would see. Since it would only need to illustrate the small area that the fovea would see, the resolution of the display could be very modest. When we look at a television image, we do not, cannot, see it all at once. Instead, we move our eyes around, taking it in, in an almost serial fashion. The impression of a whole image is constructed in the brain, although our eyes can never see it. Even though no such fully portable goggles exist now, current efforts to develop flat screen displays by many companies around the world guarantee that their appearance is just a matter of time.

With such a system the computer would have absolute authority over the viewer's visual perception. Floor, walls, and ceiling would cease to exist and what the participant would see would be completely defined by the computer. Participants could fly. They could see chasms open underneath them; they could make the firmaments respond to the wave of a hand. The real environment would have no significance. Any space could be used. Of course, the computer would have to prevent participants from walking into walls.

Holography

Holograms are film images created with lasers which store multiple three-dimensional views of a scene. As viewers move their heads, they see the image change in exactly the ways it would if the scene were real. With a projection hologram, the image appears to be floating in midair between the viewer and the film.

While there has been some effort to generate holograms by computer, the techniques required are in their infancy. Since each hologram contains many views of a scene, its generation requires much more processing than a single raster graphic image. Thus, real-time computer-generated holograms are unlikely to figure in the design of Responsive Environments for some time.

AUDITORY EXPECTATION AND DISPLAYS

The conventions of everyday auditory experience are quite different from those of vision, as sounds are less responsible for determining our sense of place. A given sound may be associated with a certain environment, but if the visual information conflicts, suggesting the viewer is elsewhere, the auditory evidence will be overruled and the visual interpretation will dominate. Perhaps this is because radio and phonographs constantly bring us sounds unrelated to our physical environment. While television, photographs, and films do this too, they are directional and exist as clearly defined, bounded perceptions. In addition, while there are static sounds like the hum of a fan or the roar of a highway, most sounds start and stop, accompanying dynamic phenomena, such as slamming a door or blowing a horn. Since sounds often occur and disappear, we must be able to process them after the fact. Our understanding of speech, in particular, relies heavily on memory. We need to remember in order to understand.

Auditory expectations are less crucial to our interpretation of the physical world than visual expectations. Thus, toying with them has less impact on our sense of place. In addition, while the computer can theoretically be used to generate any sound and has been used to synthesize almost every instrument, less effort has gone into simulating other real-world sounds. However, there is a well-defined tradition of abstract sound sequences — music. While it is not necessary that an Environment produce music, musical expectations can be used in varying the sound feedback.

After a pattern of responses has been established, it becomes expected. Subsequent responses can deviate just enough to surprise, but not so much that the participant ceases to see that a connection exists between action and response. Participants should have a sense of what the response ought to be. Their predictions should be right often enough that they trust the orderliness of the Environment, but wrong often enough that they are never absolutely certain that the expected response will reoccur.

Since we rely less on sound for orientation, we are less susceptible to manipulation by sound alone. Sound can be used to reinforce desired behavior, but this process is slower than the instant visual suggestion used in the Maze interaction. However, there are very effective techniques for accomplishing more general changes within a participant. Movie sound tracks are used to set moods and to enhance the emotional impact of visual events, indicating that it should be possible to communicate feelings of suspense, foreboding, or merriment using the same musically simple techniques.

Hardware

The needs for immediate computer-controlled response dictates the use of some kind of electronic sound synthesizer. First-generation sound synthesizers were not well suited to computer control. These analog devices were manually controlled by knobs and physical changes in the wiring. Later versions of these devices included some provision for computer control. More recently, digital synthesizers have been constructed. In 1976, I implemented one based on Fourier synthesis. Similar devices have been implemented at Bell Labs and the University of Pennsylvania.[8] In principle, these devices are capable of synthesizing any sound. However, there is an irony associated with their use. It requires considerable computer power to control them and significant disk storage to hold the control information.

Another hardware development of interest is the introduction of economical speech synthesizers which allow the computer to generate speech sounds. The most notable of these is the one used in the Texas Instruments "Speak and Spell" calculator. A computer-generated voice asks a child to spell designated words by typing on its alphabetic keyboard. After the word is typed in, it tells the child if the answer is correct or incorrect. The quality of the speech and intonation is very good, and the product is well crafted for the interaction it creates. However, the storage requirements of the system, even with a very modest vocabulary, are considerable.

Since one of the themes of the Responsive Environments has been experimentation with nonverbal communication and since the use of synthesizers for generating explicit English statements is probably too obvious to be aesthetic, a conversational Environment is undesirable. But, the idea of the computer using words or syllables to respond to footsteps or handwaving, without the intent of creating a traditional dialogue, presents interesting possibilities for creating a kind of Environmental poetry. Verbal response could also be used to create a specific atmosphere within the Environment. For example, a hostile atmosphere could be created by letting a buzz of gossip swirl around the room as the person explored it.

A word could be used as a response to the participant's footsteps. Its numbing repetition would make it into a sort of mantra. Speech sounds could be used as responses without organizing them into recognizable words. While the result would be a meaningless babble, it should be possible to impose a phrase structure on the sounds and to modulate their emotional tone so the Environment would appear to coax or warn or congratulate the participant in an unknown tongue.

The use of realistic live or recorded sound is limited in general but of interest in a few special cases. It might be desirable for the computer to have a vast repertoire of real-world sounds that it could call up at a moment's notice. Thus, the sounds of animals, city traffic, subways, airplanes, ocean, insects, or crowds could be associated with objects and events in the Environment or used to create a sense of ambience to augment or conflict with the scene portrayed on the screen. With current disk technology, a large inventory of such sounds could be stored.

A number of simple interactions could be based on this type of selection. If the switching were done rapidly, the result might be a collage of sound. Or, a number of musicians could play a piece in separate spaces, so that the sounds from each instrument could be picked up individually. The musicians would wear earphones enabling them to hear one another. As the participant moved around the Environment, activating separate speakers, the participant would hear only one instrument at a time. By moving rapidly, triggering the speakers in quick succession, a person could gain a sense of the piece as a whole.

Alternatively, the participant could be encouraged to speak or yell by programming the Environment to respond to noise. If the participant's voice were recorded on several different tape loops, or on a large disk, it could be played back on a selected fragmentary basis. Finally, movement around the darkened Environment could be used to make a real-time auditory exploration of another public place. Numerous concealed microphones or a single shotgun mike could be controlled so as to allow selective eavesdropping in the other place as the participant moved around the Environment. The relative positions of the various conversations would be preserved but all visual information would be lost. Naturally, the speakers would have to be alerted to the possibility of being overheard.

Another dimension explored in GLOWFLOW involved changing the origin of the sounds by switching among a number of speakers so the sounds moved around the room or bounced from wall to wall. With a large number of speakers and some special processing, the result can be a texture of sound evoking thoughts of wind and rustling leaves. Selecting speakers is also useful when several people are present so each person's sound response follows that person around the room physically.

TACTILE DISPLAYS

The sense of touch is not as easily addressed because there is no electronic way of projecting its stimuli. On the other hand, if the participant were to wear a suit containing a number of electrode stimulators, effective results might be achieved. Similar arrangements have been used to bring rough visual images to the blind and research is uncovering tactile illusions that make such stimulation worth exploring.[9] These electrically induced sensations could be used in conjunction with a sensing system that reported the position of the person's body. For instance, if a holographic object were projected, electrical stimulation of the hand that coincided with the projection would be interpreted as "touching" the object. As mentioned in Chapter 6, the sensation of touch can also be communicated audibly. When the participant's image touches another image, if they hear a sound, that sound becomes the "feel" of the other image.

An aspect of touch is the resistance that objects offer our efforts. If we push a wall it resists. If we push a soft chair, it yields. If we grasp an object in our hand its presence prevents our hand from closing. If we lift it, we feel its weight. One of the few systems that consider this kind of feedback is the GROPE system developed at the University of North Carolina. With GROPE, one manipulates objects within a graphic world using the kind of remote manipulator used for handling radioactive materials (Fig. 7.12). Movements of the physical manipulator controlled by the user's hand are mirrored by movements of a graphic manipulator in the display space, which also contains graphic blocks shown on a graphic table. The task is to use the manipulator to stack the blocks. The reality of the task is heightened by force feedback through the manipulator. When one attempts to move the graphic arm through the graphic table, the manipulator stops. Since the table is in the way, you cannot move the manipulator through it. When you succeed in lifting one of the graphic blocks with the graphic manipulator, the weight of the block can be felt through the real manipulator.[10]

This system is limited to a single arm and hand. However, it would be possible to design a skeleton that would

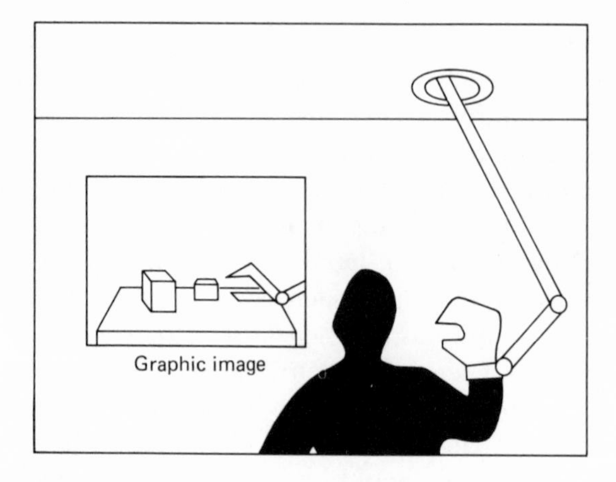

Fig. 7.12 The grope system

strap onto the participant's body that could allow or resist all possible movements of the head, limbs, torso, hands, and fingers under computer control (Fig. 7.13). With such feedback, the person could climb nonexistent mountains, swing from illusory trees, and grapple with graphic phantasms. In fact it would even be possible to place the whole apparatus in free fall and accelerate in any direction necessary to produce the desired sensations of gravity and motion.

THE OTHER SENSES

The remaining senses are difficult to reach by computer-controlled displays. Consequently, they are rather unimportant in this medium and will be dealt with only briefly. By itself, taste is extremely limited and is seldom associated with behavior other than eating. Smell is difficult to control because of the physical behavior of odor-causing chemicals. Odors move through the air slowly and linger for some time so a wind tunnel would be required to allow rapid delivery and evacuation of smells. The sense of balance and momentum can probably be stimulated visually as suggested by the Mystery Ride. All of these senses are probably best left out

Fig. 7.13 Skeletal harness

of a Responsive Environment until they can be dealt with more effectively.

Electrode Stimulation of the Brain

The ultimate display might bypass the sensory receptors and go directly to the brain. Recent experiments with blind and deaf people have demonstrated the feasibility of introducing coherent sensory information directly to the brain.[11,12] If a similar capability were added for the other senses a total synaesthesia could occur. Smell sensations might arrive in staccato fashion, in counterpoint to tastes and touch. Such possibilities are exciting and frightening. The courage to take such steps will undoubtedly be found first by those handicapped by the physical loss of these senses.

While the preceding discussion indicates a bewildering number of possible responsive displays, it also suggests that effort should be concentrated on the development of a unified computer-video medium enhanced with electronic sounds.

While still limited, such a response capability would be more than able to tax the information-processing capacity of the participant as he or she moves around the space, particularly if the responsive relationships keep changing.

NOTES

1. "John Whitney Interview Conducted by R. Brick," *Film Culture* 53, No. 7 (Spring 1972):39–83.

2. D. Weiner, "Test Tubetelevision: at WNET's Experimental Workshop," *American Film* 4 (March 1979):33–4.

3. N. Burtnyk & M. Wein, "Computer Animation of Free Form Images," *ACM-SIGGRAPH Proceedings* 9, No. 1 (1975):78–80.

4. Peter Foldes, "Hunger (Le Faim)," National Film Board of Canada, Learning Corporation of America (1974).

5. W. M. Newman & R. F. Sproull, *Principles of Interactive Computer Graphics* (McGraw-Hill, 1979).

6. F. I. Parke, "Computer Generated Animation of Faces," University of Utah Computer Science Department, UTEC-CSc-72-120 (1970), NTIS Ad-762 022. Abridged version in *Proceedings of ACM National Conference* (1972):451.

7. Ivan E. Sutherland, "A Head Mounted Three Dimensional Display," *Fall Joint Computer Conference* (AFIPS), 33–1 (1968): 757–764.

8. H. G. Alles, "Music Synthesis Using Real Time Digital Techniques," Technical Report, Bell Laboratories, Murray Hill, New Jersey.

9. The Neuroprostheses Program, "Data Processing, LSI Will Help to Bring Sight to the Blind," *Electronics* (January 24, 1974):81–86.

10. James J. Batter & Frederick P. Brooks, Jr., "GROPE-1," *IFIPS Proceedings* 71 (1972):759.

11. "An Electronic Link to the Visual Cortex May Let Blind 'See'," *Electronics* (December 20, 1973):29.

12. "C-MOS Implant to Aid Deaf," *Electronics* (February 20, 1975):38.

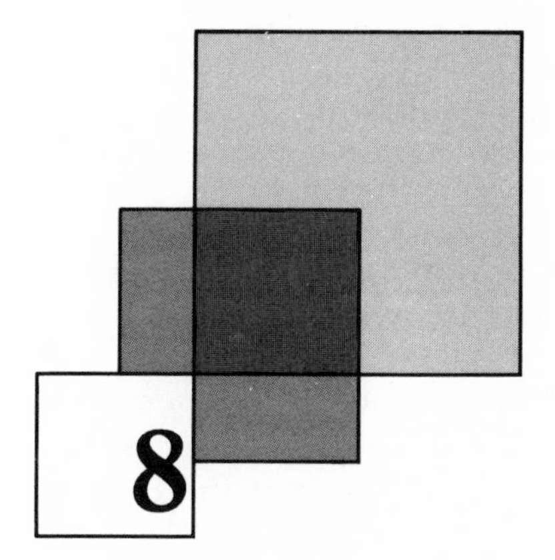

CONTROLLING THE SYSTEM

Between the sensory hardware and the auditory and graphic displays lie the hardware and software that coordinate the behavior of the system as a whole. The hardware includes the processors that analyze the input and generate the output as well as interface systems that give the computer knowledge about and control over the harsh analog world that exists outside its benign digital realm.

As previously indicated, the processing requirements of this kind of system cannot be accomplished in real time by a single processor if the scope of the interaction is going to be nontrivial. Therefore, it is necessary to consider a hierarchical system architecture that allows a number of processors to work together. There are a number of ways this can be done, each with its own strengths and weaknesses. While a number of experimental systems have been developed, the focus has been on the problem of how to create the generalized supercomputer of tomorrow, rather than on how to create a system that is organized to speed up a particular application. When this is the intent, as in the case of the Responsive Environment, it is necessary to consider the nature of the processing task and the general philosophy of its implementation to judge what architecture is most suitable.

PROCESSING TASK

Context

The general goal of the responsive system is to understand as much as possible about a participant's behavior, devise the most intelligent response, and respond with the most vivid displays feasible in real-time. Since the potential volume of input data is tremendous and the potential amount of output even larger, the computations required to totally digest the information available are beyond any processor — even the human brain.

We live in an infinitely complex reality. We are incapable of absorbing all the information potentially available to us at any random moment. Clearly, if as we walked down the street we tried to attend to the muscular activity of our legs, the movements of our diaphragms as we inhaled and exhaled, the ruminations of our gastrointestinal tract, the sensation of our feet hitting the pavement and the swinging motion of our arms, we would find ourselves unable to deal with just the information supplied by our own bodies.

Compound this information with the sounds of human voices, automobiles, birds, the sight of pavement, people, buildings, trees, and sky, and it is easy to understand Lawrence Sterne's observation in *Tristram Shandy* that it would be possible to spend an entire lifetime in an attempt to fully apprehend the events of just one day.

In order to cope with the potentially mind-boggling amount of information available to us at each moment, we exist in limited, understandable contexts. Within each context we select information that will help us to function. To cross a street successfully, it is far better to attend to the activities of automobiles than to the inner workings of our bodies.

Context is not solely determined by place. Two people can be in the same place and see it quite differently because they notice different things and interpret what they notice in terms of different goals, past experiences, and current train of thought. A passenger in an airport will notice signs for flights, gates, rest rooms, ticket counters, restaurants, news stands, and gift shops. A workman in the same airport might see only electrical outlets, trash cans, and burnt-out lights. In other words, each is looking for certain things and ignoring others. The choices are individual, but the necessity for choice is mandated by the limited real time processing power of the human brain.

We will continue to call the organizing principle that governs this selectivity a context. It subsumes the currently active filters through which an individual interprets the world and controls their responses. The context includes the physical environment, the state of the person's body and the activity or train of thought the individual is involved in. The benefit of this principle from a processing point of view is that from one moment to the next most of the information does not change. All the person has to do is verify that the context is still the same. This means that most of the processing and perceptual power of the brain can be devoted to the task at hand.

Context is not a rigid concept. It allows for change. In many cases it expects change and even predicts it. As I drive down the street, my perception is changing, but it is changing according to a set of transformations that I associate with driving. This continuity allows me to predict where objects perceived one moment will be the next and greatly simplifies the task of processing. There are also likely changes of context associated with a particular context. As I leave one room I expect to find myself in another room, an office, a corridor, a stairway, or out of doors.

Not all situations are as predictable as these two examples. There are surprises and new situations. However, human beings do not function well when they are knocked out of their current context, especially if the new situation is one for which they have no preparation. Most people avoid such experiences completely, or experiment with new contexts in tightly controlled ways. Thus, while not every situation is familiar, it is amazing how much of our experience fits within the idea of a context. If a person is suddenly thrust into a completely new or unexpected situation it takes some time to fully apprehend it.

I was once standing on the curb at an intersection as a camper with an extended rear-view mirror on its right side drove by. The mirror struck me on the side of the face and spun me around. While I was unhurt, my sense of continuity was completely severed. As I picked myself up from the ground, I saw my books and picked them up mechanically. If instead I had seen my toothbrush, I believe my first impulse would have been to brush my teeth. It was several minutes before I had reconstructed where I was, what I had been doing, and why people were asking if I was all right. I suspect that there would be a similar sense of disorientation if a person were moved while asleep and awakened in an unfamiliar place. We are very dependent on making use of continuity and maintaining an ongoing sense of context that provides significance for every perception and a framework for every action.

Context in the Responsive Environment

The Responsive Environment has only a small fraction of the processing capability of the human brain and is thus desperate for any shortcut that will rationalize its processing task. Thus, at any instant, it is assigned a context that may be a single visually displayed space consisting of a scene, the participant's image, several objects, and the fact that the participant's movements generate sound feedback. As the participant moves about the room, the perspective transformations in the displayed scene and the generation of sound responses are part of the current context that is expected to be maintained from one second to the next. Only certain aspects of the participant's behavior are relevant to this framework and most of the display will not change on a moment-by-moment basis.

If a change does occur, it will be derived from the previous display through simple transformations. Thus, at any point in time, the input is being analyzed only with respect to very specific criteria, condensing the large amount of potential data into a much smaller amount of symbolic information. On the basis of this condensed information some rather simple decisions are made and a brief description of the intended response is created. This information is passed on to lower-level processors which take the small semantic description and generate full auditory and visual displays (Fig. 8.1).

Even with the simplification that the context idea provides, the processing task is far beyond the capability of any single processor. While we are accustomed to thinking of computers as fast, there is actually very little they can accomplish in a small fraction of a second. Therefore, it is natural to think of using a number of processors. Certainly, each of the processes discussed earlier would require one or more computers to execute it.

There are many ways that computers can be tied together. Many of these are motivated by concerns other than speed. To appreciate these options on an intuitive level, consider the ways that we have of organizing groups of people in cooperative effort. One such model is the committee that

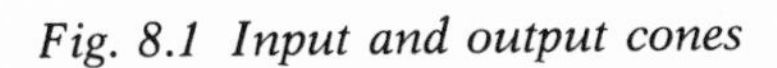
Fig. 8.1 Input and output cones

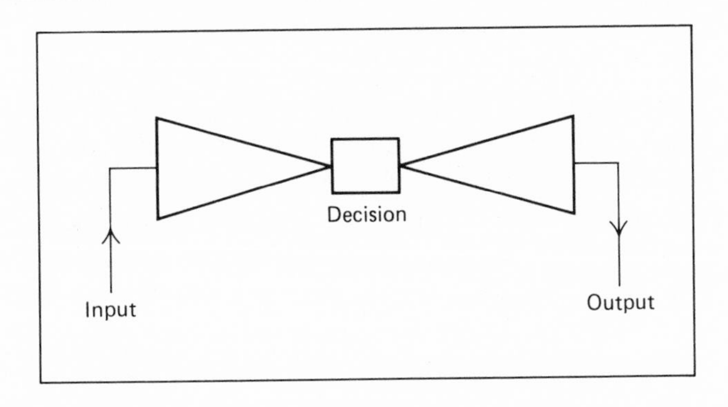

is loosely organized and not noted for its speed or productivity. Perhaps surprisingly, much academic research in parallel processing is enamored of this approach. The reason is that decentralized control promises greater reliability. It is useful where the problem to be solved is ill-structured and the flow of control unpredictable. Also, such systems are considered more intellectually interesting than those with centralized control. A second approach is a hierarchy. Information flows up from the bottom, command and control filter down. Whether this is a good model for getting things done depends on the degree to which higher levels delegate decision-making authority to the lower ones. Where there is no delegation, the result is a bureaucracy which is a fair description of most traditional operating systems.

Perhaps the most productive organization is the assembly line. The work passes from one worker to the next with each bringing it one step closer to completion. All workers are always busy and work at the same pace. The assembly line, which computer scientists refer to as a pipeline, is very relevant to the reflex level of response processing. However, as the reflex cone suggests, there is a compression of the input and an expansion of data towards the output. The result is that a number of parallel pipelines can operate on the input. These will coalesce as the input is processed. When the response decision is made, the output generation will diverge into a number of parallel pipelines (Fig. 8.2).

While the number of processors is significant, the kind of processors may be more so. The entire history of computer design has been guided by a single set of principles which are only now starting to yield to new ones.

To understand the limitations of traditional computer architecture, it is useful to think of the computer as a factory with thousands of workers. However, unlike any rational factory, the computer has only one workbench and one set of tools. Only one worker can function at a time. To do this, each worker must go to the workbench, perform a simple operation, and get out of the way. The work can be left on the bench if the next worker is going to use it. If not, it must be stored somewhere. Even the most militant union would agree that this is a ridiculous way to run a factory.

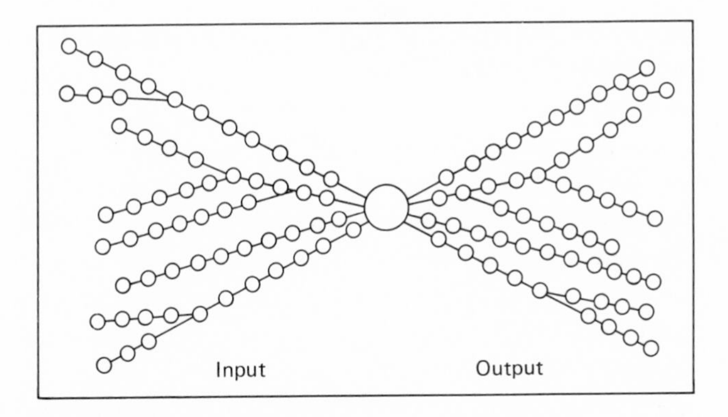

Fig. 8.2 Pipelines in reflex cones

To get a feeling for what this mode of operation implies, imagine a teacher calling roll in class. Rather than just reading the list and having the appropriate student respond, the teacher would take each name on the list and ask each student if that was their name. Although most research in parallel processing has consisted of tying together a number of processors of this type, the greatest speed improvements are gained from specialized processors and those that are inherently parallel in operation. While such an approach may seem less flexible, a quick look at the anatomy of the human brain reveals many specialized processing structures. The retina of the eye performs an enormous amount of computation before it sends anything to the brain. A Responsive Environment faces a very similar processing requirement. Therefore, it is to be expected that specialized processors will be needed to make the system operate in real time. It is almost certainly true that drivers close to either the input or the output should be specialized to their task.

SYSTEM ARCHITECTURE

The control of a Responsive Environment involves at least two levels of function which require quite different architectures and modes of operation. At the bottom level is the reflex processing which controls the real-time execution of the system within the current context. Above this is a level of cerebral processors which are not involved in the real-time response, but which are responsible for the character of the experience and the aesthetic goals of the interaction. In between are the hardware and software systems which allow the slow intelligence processing to assert overall control over the reflex processing without getting in the way.

The demands on the processors involved in the reflex path are total. They must be dedicated to the task of performing their step in the processing. There is no time for them to have any larger awareness of what is going on.

At the lowest level in the hierarchy are the drivers which operate very close to either the input or the outputs. Feature detectors working on the input and graphic element generators producing output are examples. These processors are dedicated to a particular function, which they perform continually, passing any results up to higher level processors. Each of these processors is looking for very specific things. This function may be changed by a superior processor which instructs them to execute a different portion of their code or loads a new set of code into them. There may be several layers of such processors. They are distinguished by being dedicated to either input or output, but not to both.

The reflex processors operate above the drivers. They determine if events are within the current context. If they are, these processors choose the appropriate response and send the commands that effect it to the output generators. The response may be the movement of the participant's symbol as in the Maze program. Or, it may be a change in the displayed perspective or the response of a mathematical model to a parameter controlled by some aspect of the participant's behavior or the behavior of a simulated computer creature. If the participant's behavior triggers a change in context, the reflex processor directs all subordinate processors to change their operation. It also notifies higher-level processors to fully stage the new context. The reflex processors are distinguished by the fact that they make real-time response decisions and they deal with both input and output information. Several reflex processors may exist in parallel or in a hierarchy.

In between the cerebral processors and the reflex processors is the staging system which loads the software and data associated with adjacent contexts into a staging memory. When a context switch occurs, it transfers this information to all affected processors as rapidly as possible. This subsystem has control of a large disk file system and high-speed access to all processors.

The support software at the staging level is essentially an anticipatory operating system. Usually, an operating system waits for an event to occur before it starts to cope with it or take advantage of it. In this case, the staging system is only minimally interested in what needs to be done now as it has always anticipated and prepared for immediate events. It is more interested in getting ready for what might happen next. Thus, it will stage the code for every immediately possible path, even though only one path will actually be taken.

At the top level are the more cerebral functions that have time to observe what is going on in a detached way and can devote themselves to the larger perspective of the composition as a whole. They must take into account:

1. The current context.

2. The history of the interaction so far.

3. The adjacent contexts.

4. The planning of transitions to one of the adjacent contexts.

5. The need to realize immediate goals of the interaction, such as slowing the participant's actions or inducing movement to a new location.

6. Other constraints that may be part of the composition, such as a time limit in a given context.

The cerebral processors are distinguished by the fact that they are not directly engaged in real-time processing. Their responsibilities lie in monitoring the interaction and perhaps redirecting it. They ask, "What kind of participant is this? Has this person been in the current context too long? Or should the current context be enriched, some new ingredient or entity introduced?" None of these deliberations need be accomplished instantaneously. They may require a few seconds. Once a decision is made, it may be enacted immediately or it may simply change the future path of the interaction.

The composition processor would monitor the interaction and consider it in terms of the original composition. It will contain a schedule of events that are to occur within the context and an amount of time allocated to each. These may be changes in the focus of the system's perception, changes in the output response, or changes in the conditions that will cause transition to a new context. For instance, a minute was allowed for participants to absorb the idea that the symbol on the screen would track their movements on the floor in PSYCHIC SPACE. The last event in the context schedule is always a default transition which occurs after a specified amount of time if none of the conditional transitions has been triggered. In addition, the character of the interaction may be monitored in light of high-level criteria associated with the current context. Based on information such as rate of movement, the composition process may decide to add or delete a feature of the context. As technology improves, it will be possible to make the judgments of this processor increasingly intelligent. To date, however, this processor has mainly had a scheduling and monitoring responsibility. It is where intelligence processing would go were it available.

Another important cerebral processor is the off-line large-scale processor that is used for software and algorithm development. This machine should include high-level languages both for systems programming and for Artificial Intelligence efforts that would allow the system to learn from past experiences while no one was in the Environment. This Artificial Intelligence connection is just beginning to be developed.

The connections implied so far fairly represent the logic and control flow of the system; however, there are occasions where it will be desirable to short-circuit the response chain to provide data from the input end of the chain directly to a processor at the output end or vice versa. One motivation for doing this would be the desire to use the participant's image as a graphic element in the output. This data could be sent through the length of the response chain, but given its volume, it is reasonable to provide a separate data path for it, especially since it is available on the same video bus that the graphics processors are plugged into.

Another example of such a short circuit would be the situation where the system wishes to test the intersection of the participant's outline with a computer-defined graphic image. If the graphic is treated as just another input image, the sensing hardware can take care of digitizing the intersection. In fact, the contact of two graphic objects could also be tested for in this way. Therefore, the reflex cones described earlier must be amended somewhat to include both the cerebral processors and these short-circuiting connections (Fig. 8.3).

A powerful bias has emerged in the discussion of hardware architecture. The traditional casual definitions of real-time processing are too lax. The normal concept of an op-

Fig. 8.3 Conceptual representation of system architecture

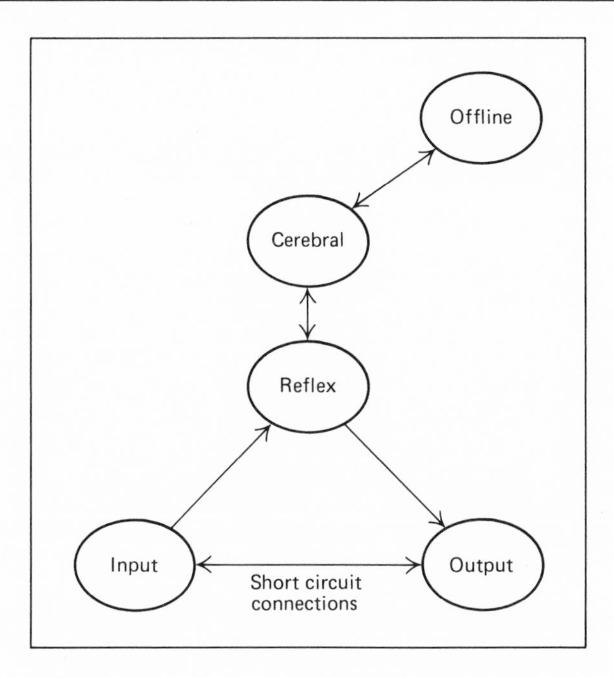

erating system is banished. This application requires operating software which does its bidding immediately, not a separate entity of which requests are made.

Context

Thus far, we have assumed that the contexts within a composition are joined in a network (Fig. 8.4). The current context is a node that is directly connected to a small number of adjacent nodes, each of which is in turn connected to a few others. Associated with each node are procedures for initialization which govern its operation, detect its leaving, and clean up after it is exited.

In a broader sense, a context is a framework into which new elements can be injected or the presence of an intelligent overseer inserted. Similarly, the net of interconnected nodes need not be fixed. The overseer can consider the current state of the interaction and overrule the current script, jumping

Fig. 8.4 Response contexts

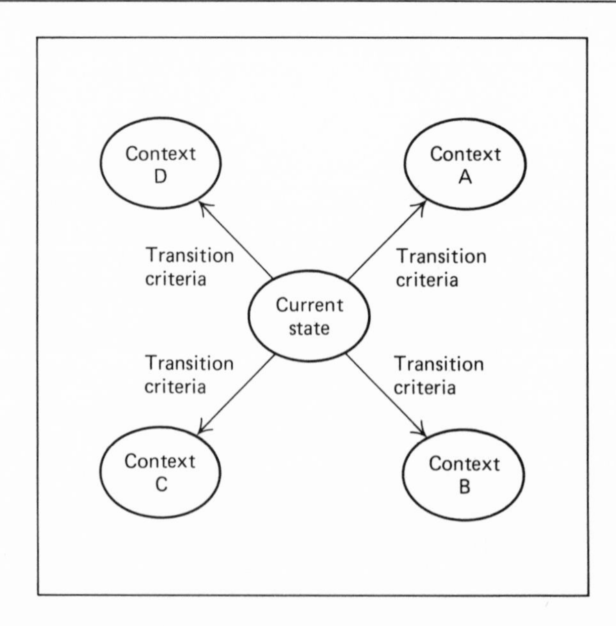

from one situation to one completely unrelated but in some way more appropriate.

One occasion when this might occur is when the system discovers some pathological condition, such as the very rapid transitioning from one context to another and then another, or back and forth between two adjacent contexts, or the disappearance of the participant, or the apparent creation of a new participant. Any of these conditions might signal the presence of a jokester, someone who has psyched out the system and learned how to confuse it. While this is not exactly a capital crime, there is no reason the Environment should not be aware of this possibility and be equipped with special compositions to play another game on the jokester's game. More will be said about Artificial Intelligence control for the Environments in Chapter 10.

SOFTWARE TOOLS

In addition to the software that actually controls the Environment in real time, there are a legion of general software tools required to develop the specific interactions. While the software that provides real-time control cannot be overly complex, this off-line software can be sophisticated as time is not a constraint on its operation.

At the lowest level are the exercise programs that are used to investigate the behavior of an input or output device. In the case of an output device, there is a need to generate experimental output to determine effective building blocks from which to develop higher level responses. All of the synthesizer options mentioned in Chapter 7 had one feature in common; they were controlled by waveforms. Therefore, the composer of an interaction using the synthesizer must experiment with different waveforms in order to be able to associate the combinations of waveforms that define a single complex sound.

There is a similar need to experiment with graphic capabilities. Graphic objects and scenes must be introduced to the computer through a powerful interactive graphics system based on a data tablet. This system should be capable of designing animation sequences and testing the rules of articulation and behavior of graphically defined creatures. Both animation by key frame interpolation and by modeled behavior should be available. It should also be possible to create graphic Environments and move about them using the tablet to simulate the behavior of the participant.

It is also very useful to save input provided by a participant during an interaction. This data is used to test pattern recognition algorithms and the higher-level software that controls responses. The availability of such tools allows the artist to compose experiences and see how they work in realistic situations. Ultimately, the artist will want to refine experiences in ways analogous to a painter reworking a painting. To do this, the artist will temporarily assume the role of participant, moving about the Environment, taking alternate paths, and exploring the final result. It would be desirable for the artist to have the same ability to manipulate the composition during this process as would be available to an editor manipulating a text. Obviously, this would require a method of communicating explicit commands to the computer from the interactive space. Methods of achieving this will be addressed in Chapter 12.

These software tools deal with the system as a self-contained unit. However, this system can be used in conjunction with other machines. For instance, the Environment can be used to run behavioral studies. Data taken during an experiment would be stored on the system disk as an inverse of the usual staging operation. Later, this data would be shipped to a large-scale number cruncher that would do the statistical analysis.

A large machine would also be useful for simulation of new hardware designs and firmware algorithms. The availability of high-level languages would also make the use of learning algorithms feasible. Currently, such Artificial Intelligence processing is an off-line development effort. Later, it will become a standard part of the system operation even if it is actually run on an external processor. A record of recent interactions will be available to the system if there is no one in the room. When there is nothing else to do, it can ruminate about its experiences and try to learn from them. This will be the computer equivalent of dreaming. Whether the result is better performance in some objective sense is not necessarily the most interesting question. If the system learns over a long period of time and considers its experiences from a number of perspectives, the fact that it chooses to change its behavior in a coherent way will be as interesting as its ability to optimize a specific skill would be.

These hardware and software systems represent a long-term effort to develop the tools required to create and to compose within the new medium. Emphasis is currently on hardware because the needs of real-time systems are poorly understood. Therefore, much work must be done to build a system that functions at the speeds required. There is also a need to coordinate the capability for reflex behavior with a greater intelligence that operates at slower speeds. As this goal is achieved, emphasis will shift from hardware to compositional software. As the awareness and reflective powers of the system evolve, it should be possible for the artist to represent wit and intelligence through the system as well as in any other medium. In the distant future, the artist's task in this medium will change. The goal will be to raise an aesthetic intelligence whose life work is artful interaction with human beings.

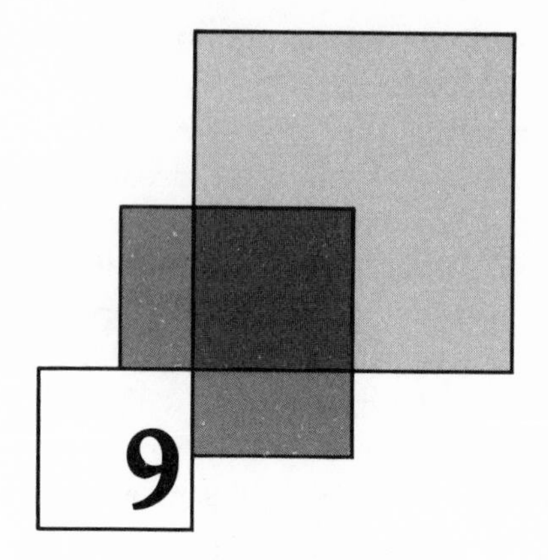

VIDEOPLACE

GLOWFLOW, METAPLAY, and PSYCHIC SPACE were separate events implemented with very different technology and directed towards different goals. Reflection on these shows and consideration of the technical issues detailed in Chapters 6–8 led to a conceptual synthesis. The new conception is based on the premise that two-way telecommunication between two places creates a third place consisting of the information that is available to both communicating parties simultaneously. When two-way video is used, a shared visual environment can be created. We call this the VIDEOPLACE. The sense of place can be embellished by a computer until a complete artificial reality is created that the participants perceive through video projection — and experience through the participation of their video image in the portrayed world. Although it was conceived in the context of two-way communication, VIDEOPLACE can also be used to stage interactions involving a single participant.

This concept of a communication space separate from existing physical spaces has provided a continuing theme for aesthetic thinking and a focus for hardware development.

VIDEOPLACE improves on the earlier Responsive Environments in that it combines sensory and display systems into a single integrated video system architecture. In addition, it is conceived as a flexible medium rather than as a single exhibit. The VIDEOPLACE concepts and components are rich enough to support a wide variety of compositions that will take years to exhaust. Thus, as VIDEOPLACE techniques are perfected, developmental emphasis will shift from hardware to software to aesthetic composition.

VIDEOPLACE — ORIGINAL INSIGHT

VIDEOPLACE had its origins in an incident that occurred while my colleagues and I were developing the digital and video communications for METAPLAY in 1970. I was sending digital signals from the PDP-12 computer in the gallery to a friend at the Adage display computer in the University Computer Center. Both computers had a graphic capability that enabled the PDP-12 to display the waveform being sent and the Adage to display the information being received.

At first we talked over the phone about the displays we each had in front of us. However, after a few minutes of frustrating discussion, we realized that we had a far more powerful means of communication available. Using the two-way video link described in METAPLAY, we turned the gallery camera on the PDP-12 screen. The Computer Center camera was already aimed at the Adage. Both of us could now see a composite image juxtaposing the information being sent with that being received (Fig. 9.1). We discussed the two signals and speculated about the nature and source of our transmission errors. As we did this, we used our hands to point to various features on the composite display. It was exactly as if we were sitting together at a table with a piece of paper between us.

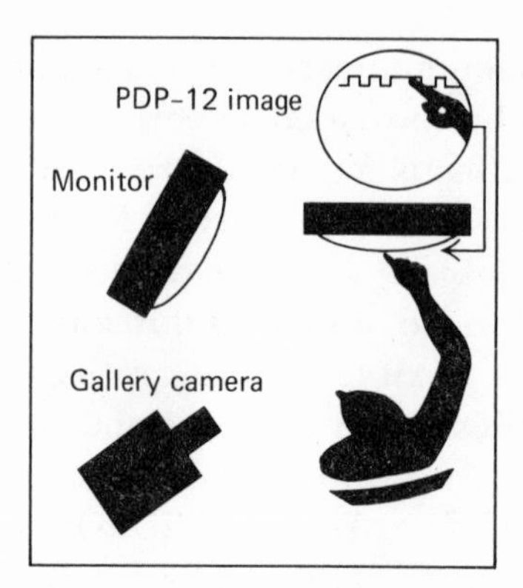

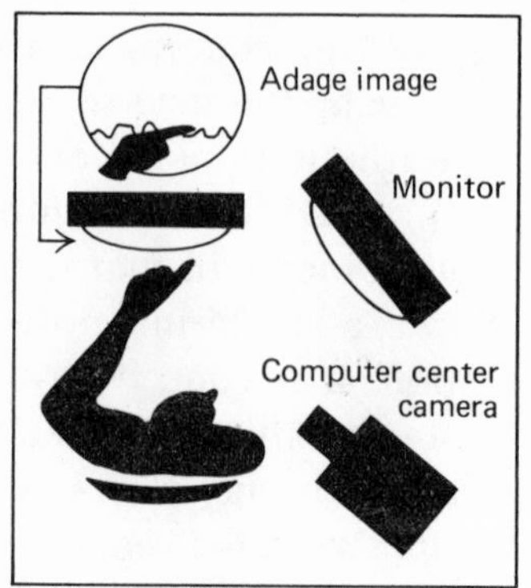

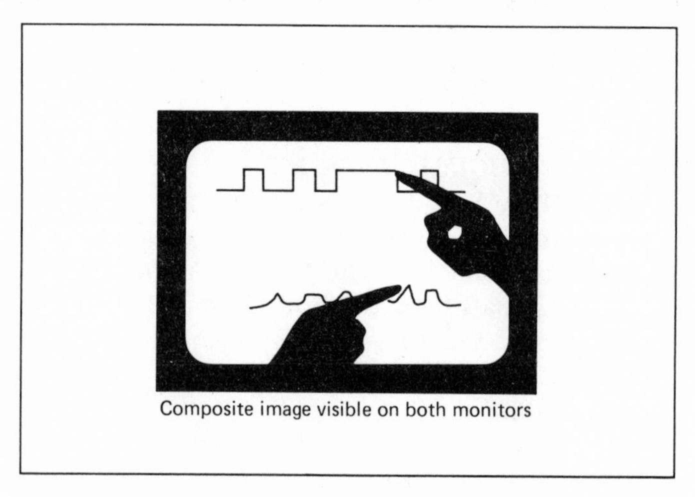

Fig. 9.1 Video communications link

After a while, I realized that I was seeing more than an illusion. As I moved my hand to point to the data my friend had just sent, the image of my hand briefly overlapped the image of his. He moved his hand. While I noticed this, its significance did not really register. However, when it happened again I was struck with the thought that he was uncomfortable about the image of my hand touching the image of his. Without saying anything, I subtly tested my hypothesis. Sure enough, as I moved the image of my hand towards his, he repeatedly, but unconsciously, moved the image of his hand to avoid contact. Even though his reactions were exaggerated and even bizarre, he never noticed my actions or his. The inescapable conclusion was that the same etiquette of personal space and avoidance of touching that exists in the real world was operating at that moment in this purely visual experience.

VIDEOTOUCH

We experimented with the sense of video touch after exhibition hours and confirmed that indeed there was a powerful effect operating. In 1972 a proposal titled VIDEOTOUCH was submitted to the National Endowment for the Arts. The piece was to consist of two Environments, each containing a rear-screen video projection of a composite image of two participants. As a single participant entered each of the two separate Environments, each screen would display both of their video images (Fig. 9.2). If their images chanced to touch, sound would be generated. The nature of the sounds was to depend on where the images touched and the size of the area of overlap. It was to be at least theoretically possible to perform a duet by maneuvering the images and controlling the ways in which they touched. Thus, two strangers would be placed in a situation where the normal embarrassment about touching was in conflict with their desire to explore this totally unexpected way of interacting.

VIDEOPLACE — THE MEDIUM

Over the next two years the concept of VIDEOTOUCH matured as its broader implications were realized. A new concept evolved — VIDEOPLACE. The experience during METAPLAY had shown that two people's images could work as if the people were together and that people felt at least some of the same self-consciousness about their images that they feel about their bodies. Later experimentation convinced me that people have a very proprietary feeling towards their

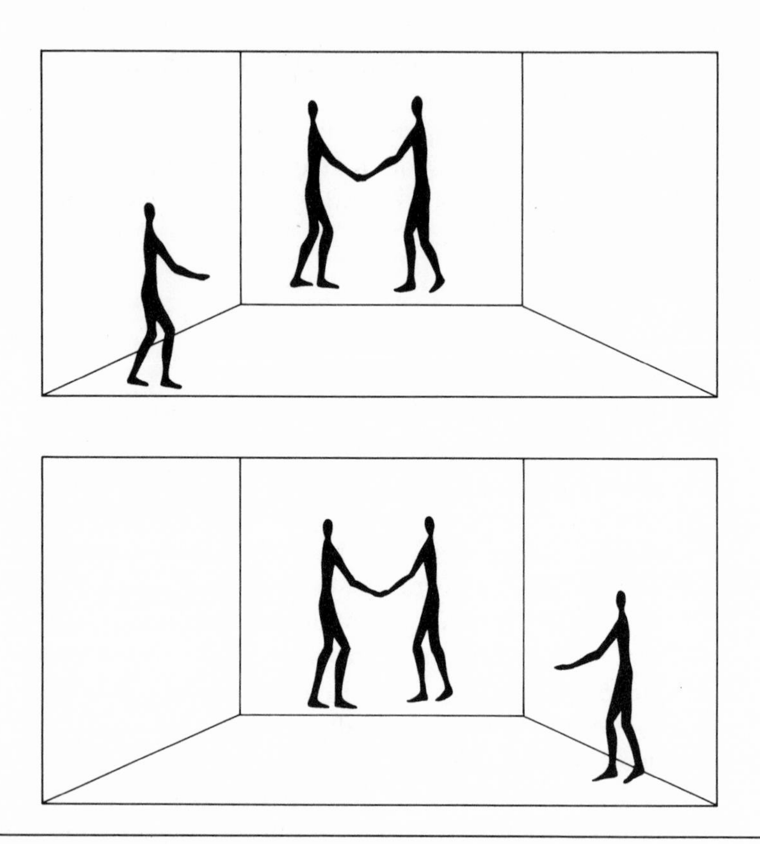

Fig. 9.2 The computer detects contact between the video images of two participants

image. What happens to it happens to them. What touches it, they feel. When another person is encountered, a new kind of social situation is created, where reality itself is in doubt. By suggesting but not duplicating familiar reality, this new encounter highlights assumptions and expectations that we are never aware of because it does not occur to us that our world could be other than it is.

The idea of a video "place" led me to consider the characteristics of a real place and how they could be reproduced or replaced by alternatives in a video Environment. If two people are together, they can see, hear, and touch each other. They can move about the same space seeing the same objects, the same scene. They can manipulate the objects present and hand them to each other. Finally, they can share an activity, such as dining together.

Familiar telecommunication devices severely limit the richness of the "teleplace." The telephone denies all of our senses except hearing. Even Picturephone, as once offered by AT&T, provided only a shaving-mirror view of the other person. Its small screen and limited field gave little sense of physical presence. It completely precluded a common visual environment or the ability to touch, and the scale was far too small for active involvement.

When you enter the VIDEOPLACE Environment, on the other hand, you are immediately confronted by a life-size video image of yourself whose movements correspond to yours. The sense of a separate reality is augmented by graphics that define a three-dimensional space that changes according to the rules of perspective as the participant moves about in it. Other effects also mirror the reality we are familiar with. For example, if you walk behind the graphic image of a door, your own image is occluded, making the existence of the door believable (Fig. 9.3). In addition, the computer can furnish the space with graphic objects and inhabit it with animated organisms.

Since the computer also has the ability to move computer-generated objects and live images around the screen, and to tell when a live image comes in contact with an object or another live image, it is possible for the computer to coordinate the movement of a graphic image with the participant's movement so that the video alter ego is able to lift, push, or throw an object or creature across the display screen (Fig. 9.4). Thus, it would be possible for a participant in New York to pick up an object and toss it to a Californian (Fig. 9.5).

However, the world VIDEOPLACE simulates need not be a real one. Unlike the real world, VIDEOPLACE is not governed by immutable physical laws. Gravity may control the participant's physical body, but not the person's image, which can float freely about the screen. Any set of movements of the body may be mapped onto a completely different set

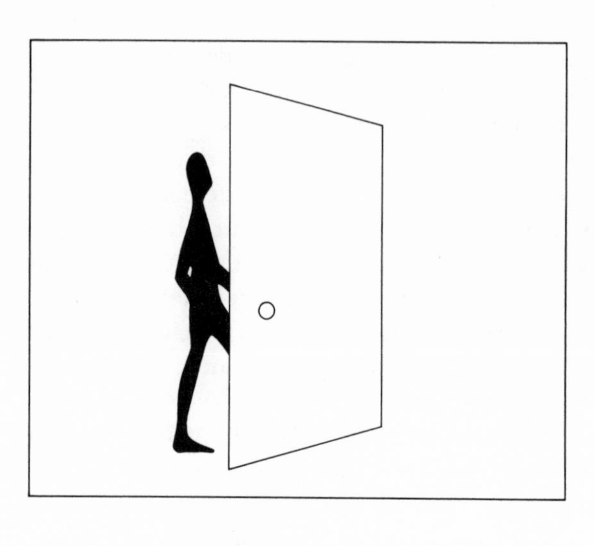

Fig. 9.3 Occlusion by a graphic object

of movements of the image in the displayed space. For example, a normal step in the real environment can become a hundred-foot leap in the VIDEOPLACE Environment, or simulated swimming motions can cause the participant's image to swim around the screen (Fig. 9.6).

Movement around the room can have an additional effect besides navigating the displayed scene. It can change it. Instead of displaying a mountain and waiting for the participant to figure out how to climb it, the mountain can grow under the participant's feet as they walk. Similarly, stamping one's feet can cause flowers to grow or volcanoes to erupt.

Fig. 9.4 The computer detects contact between the participant's image and a generated object

Fig. 9.5 Three-way catch in VIDEOPLACE

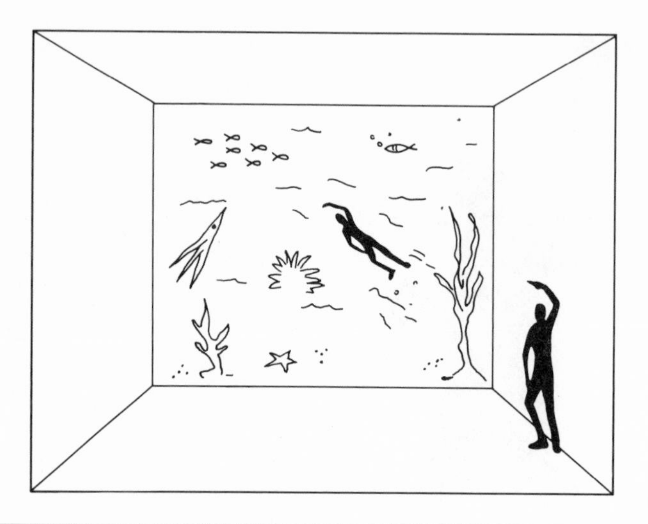

Fig. 9.6 Swimming in the VIDEOPLACE

Although VIDEOPLACE cannot literally duplicate the fullness of the real world, it attempts to invent a new model of reality with methods of interacting that are equally satisfying. Thus, while some aspects of reality are abridged in the VIDEOPLACE, others are enhanced since many of the constraints and limitations of reality can be overcome. Interactions within the Environment are based on a quest to understand the rules that govern this new universe. Expectations can be teased, leading to a startled awareness of previously unquestioned assumptions, much like the experience one has when viewing a Magritte painting.

THE VIDEOPLACE SYSTEM

A VIDEOPLACE Environment is a 16′ × 24′ room with one wall used for rear-screen video projection. A camera near the side of the screen picks up the participant's image. This image is fed to an outline sensor that digitizes the outline of the participant. The image interpreting processors analyze this outline both in absolute terms (e.g., posture; rate of movement) and in terms of the current interaction (for example, is the participant touching a particular object on the screen? has their image reached the graphic door?). Posture can be inferred from the outline itself and motion can be determined by comparing the current outline to the one before. Outline sensing is facilitated by the fact that the walls are a neutral color making the participant's image easy to find.

When the participant's actions are understood, they are reported to the control processor which decides what the response to that behavior should be. Depending on the behavior, it decides whether to move an object, change its color, move the participant's image, make a sound, etc. When the appropriate responses have been determined, the response processors in charge of graphics and sound are directed to generate them. The graphic response can involve the generation and animation of graphic objects and the manipulation of the participant's image by scan modulation techniques. The separate image elements are merged to create a composite image by an image combiner which decides which elements are in front and which are occluded by the others. The composite image is then projected before the participant.

Two or more Environments can be linked, with the addition of a little extra hardware. Each participant's image must be analyzed separately and then compared to the others and to the objects on the screen. In addition, each input image requires separate scan modulation hardware. While there are more complex decisions to be made, the basic operation of the system is unaffected.

VIDEOPLACE SCENARIO

Identity Quest

In one possible scenario, Roland, a participant, enters the VIDEOPLACE. His live image and a drawing of a knee-high door appear on the screen. He studies his new large-screen movie star status with appreciation. Then, he takes several tentative steps towards the tiny door. With each step, his image shrinks until he is as small as the door. Prompted by curiosity, he touches it. It opens and he walks through. That act places him in a new graphic space.

In the new space both the door and his image have disappeared. The only object in the scene is an amoebic blob. He scratches his head and the top of the blob ripples. He shifts his weight from his left to his right foot and the blob rocks back and forth. He takes a few steps to the right and the blob moves too. Slowly it dawns on him. The blob is he!

He jumps up and down and the blob bobs up and down — its underside slightly flattened on impact. He jumps repeatedly and notes that the blob is being flattened and its mass elongated with each jump. Moments later, elongated and transformed, the blob resembles a fish. Roland transverses the room, causing the creature to undulate across the screen in graceless parody of his motion.

Tired of walking back and forth, Roland begins to jog while the fish struggles to keep pace. Halfway across the screen, a diagonal line appears which the fish is unable to cross. Roland guesses that the line represents land because of the fish's difficulties. However, he keeps slowly walking in the same direction. The fish struggles to follow. Its struggles are finally rewarded by the appearance of two stick-like appendages on its lower body. With these, it crawls onto land. Roland continues alternately walking and jogging back and forth across the room while his strange two-legged alter ego dutifully follows. However, the motion seems to take its toll on the creature which becomes thinner and thinner until it resembles a horizontal stick with two insect-like legs.

A simple drawing of a tree appears on the screen. As Roland continues walking, it becomes apparent that the tree is blocking the creature's path. Its legs move, as Roland's do, but it walks in place unable to pass in front or in back of the tree. Moments pass with Roland and the creature motionless. Then, a second pair of appendages grow near the creature's front end. It occurs to Roland that he might try climbing the tree. He acts out an exaggerated version of climbing, raising one arm and its opposing knee and the other arm and knee. His stick counterpart moves up the tree and disappears into the foliage. For a moment, Roland's actions seem to have no effect. Then, he steps to one side and sees his own puzzled face peering out of the leaves. He moves again and his face disappears, only to reappear a moment later in another part of the tree. It disappears again, this time for a prolonged interval. Roland begins to get impatient and moves abruptly. The creature, transformed into a stick man, falls out of the tree. Upon impact with the ground, Roland's face bounces off, leaving the stick man with a round blank face. Roland moves and the stick man picks himself up.

Roland walks slowly and the tree recedes. In the distance, he spots a small figure. As he gets closer, it becomes clear that the figure is a static version of his own image seen from behind. He makes the stick man tap his real image on the shoulder. On contact, the real image splits in half and reveals another stick man inside. As this new creature turns around, Roland sees that it differs from his companion in one significant way. Instead of an empty circle for a head, it has a cartoon face. The new figure tags the old stick man, who promptly disappears.

When Roland moves again, he discovers that he is in control of the new figure. Oddly, he not only controls the movement of the limbs, but also the expressions on its face which flit from smile to frown to eye rolling as he walks. He explores the relationship between walking and facial expressions until he notices a large bulky object on the screen. As he moves his linear image towards it, he is aware that it is a huge head. Its face is his own.

It seems reasonable that since it is his face, it should be on his body. But, he does not know how to get it there. He makes the stick man push against it. It moves slightly to the side. He tries to rock it from side to side, but nothing happens. Finally he puts his hands on the floor as if he were reaching under it. It works. He raises his arms and the stick man lifts the face high over its head and then lowers it into place on its shoulders. It covers the cartoon head and shrinks to an appropriate size for the stick man's body.

When Roland moves again, his face jiggles. Its shape distorts as if it were drawn on a balloon filled with water. Roland discovers that by squeezing his hands up against his own head, he can hold his video face together. After a while, his face stabilizes.

He takes a step to the side. A frozen image of himself remains behind, as his live image follows his movement. As he takes another step, a second frozen image is created. This happens again and again until there are eight copies of his image frozen in different postures across the screen. When he takes the ninth step, a clone is created. Unlike the copies, it follows his movements. Its appearance coincides with the disappearance of the original copy. He begins to walk back across the room in the other direction. Each step yields a new clone and the disappearance of a frozen image. Quickly, he becomes the leader of a parade of selves. After eight steps, he has eight images moving in unison. He raises his arm and they all raise theirs. He kicks his leg and the video Rockettes kick as well.

Suddenly, one of the clones starts to grow while the others begin to shrink. Soon it towers over the small figures who cease to copy Roland's movements. Accompanied by a cacophony of buzzing, they attack the giant. Some cling to his feet, while others climb his body. He jumps away. One falls off. He steps on it, leaving a spot on the floor. He shakes his body. Others fall off. He steps on several, causing them to disappear, leaving more spots. He chases the others, who scurry off the screen leaving him and the spots alone on the screen. As he moves again, even the spots run off the screen, except for one. He approaches it and he realizes that it has become a hole. He peers into it and sees the tiny clones have assembled their bodies to spell "THE END" (Fig. 9.7).

This scenario is ambitious. However, from a hardware point of view it is feasible. Circuits exist or can now be designed that will accomplish each of the effects described. The software to recognize each of the events and effect each

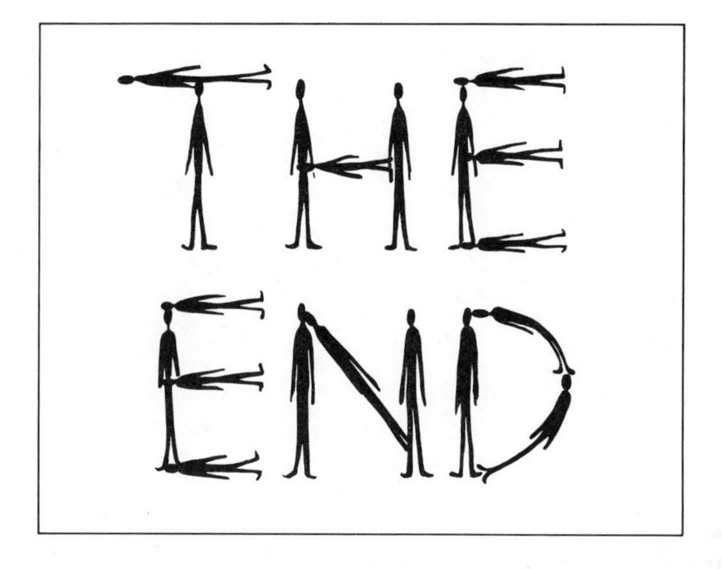

Fig. 9.7 Configuration of clones spelling "THE END"

of the computer responses can be implemented. The problem lies not in making the experience work at all, but in getting it to run smoothly. In this scenario, the participant's actions were consistent with the system's expectations and intentions. Had Roland lost his balance or decided not to cooperate with the experience, it is not clear what the system would do or, indeed, what it should do.

It is important to note that this scenario is just one example of VIDEOPLACE just as *Wuthering Heights* is one example of the novel. VIDEOPLACE is not a single piece, but a powerful medium that can be used to compose a wide variety of interactions. The scenes it portrays may be realistic or the stuff of fantasy (Fig. 9.8). The relationship between the participant's actions and the computer's response is completely programmable. In principle, our imagination is the only constraint. In the long run, Responsive Environments promise a whole new realm of artificial human experience in an artificial reality where the laws of cause and effect are composed by the artist.

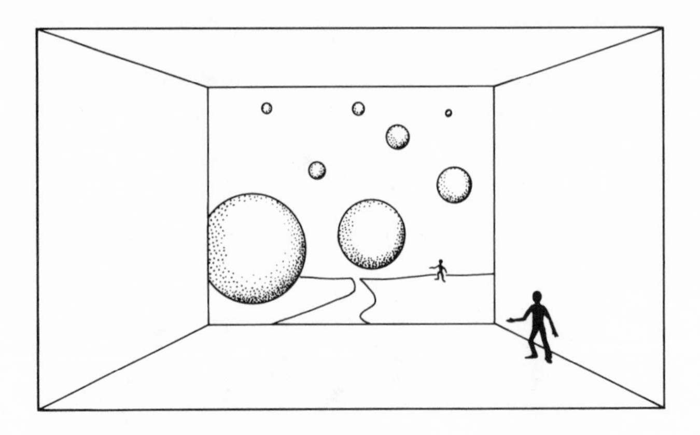

Fig. 9.8 Fantasy world portrayed in VIDEOPLACE

VIDEOPLACE: THE BICENTENNIAL EXPERIENCE

In July of 1974 it occurred to me that VIDEOPLACE would provide an ideal theme for the Bicentennial celebration. The telephone had been a Centennial project a hundred years earlier and it seemed that VIDEOPLACE would be a natural updating of the communications theme.

The thought was that there would be a single VIDEOPLACE created in different locations using satellite and microwave communications. People could then enter it from different parts of the country. All of these people could be shown together in a single composite image at all times. Or, as people moved around their individual spaces, they would encounter other people who were wandering around their own spaces. A natural greeting in this situation would be "Hello, where are you?"

Another variation would be to have a single collage of faces of people from all over the country, in which new faces would be added continually (Fig. 9.9). The result would be the realization of Andy Warhol's prediction that "in the future, everyone will be famous for fifteen minutes."

Alternatively, the VIDEOPLACE could be used to hold panel discussions among participants who lived in different

Fig. 9.9 Bicentennial mosaic

states. Or, dance pieces could be planned where the dancers were thousands of miles apart and the choreography existed only on the video screen. Note that this would be pure video; there would be no place one could go to "really" see it. In addition to these more contrived and formal events, the VIDEOPLACE would also provide a facility for ongoing experimentation by serious artists and for serious children of all ages exploring a new kind of play.

While the project was not implemented on this global scale, it stands as a proposal for the future. A more modest effort, including an exhibit at the Milwaukee Art Museum, was funded by the National Endowment for the Arts and the Wisconsin Arts Board.

In addition, the University of Connecticut Research Foundation has provided grant money for research in human-machine interaction with matching funds from the Electrical Engineering and Computer Science Department, so an ongoing VIDEOPLACE facility now exists.

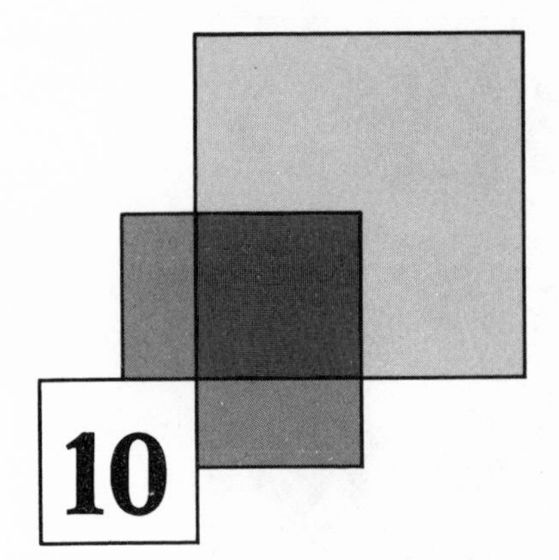

SELECTED PRACTICAL APPLICATIONS

Cybernetic systems pervade all the sciences; however, there are several fields in which the techniques described earlier will find their most fruitful application.

Since the Responsive Environment is an expression of computer technology, it can be expected to reflect and influence the field of computer science. In addition, the Responsive Environment is a powerful tool for eliciting and manipulating human behavior in both therapeutic and instructional ways in the areas of psychology, psychiatry, and education. In each of these domains it is the ability of the computer to control stimuli and define dynamic, complex feedback relationships that makes its application attractive.

COMPUTER SCIENCE APPLICATIONS

In computer science there are several areas which may be influenced by the ideas and techniques developed for the

Responsive Environments. These include the human-machine interface, real-time simulation, and Artificial Intelligence.

Human-Machine Interaction

Two basic questions underlying the development of the Responsive Environment are: what are the ways in which people and machines can interact and which of these are aesthetically most desirable? The first question is seldom asked in its full generality. At most, it is handled descriptively in discussions of existing interactive techniques. The techniques that exist have been developed ad hoc, dictated by the requirements of the problem at hand, rather than systematic study. Current communication with the computer is typically limited to keyboard, light pen, data tablet, joystick, and track ball. The Environment, on the other hand, invites a more general exploration of human-machine interaction through new forms of communication via voice and physical movements involving the whole body.

Traditional output devices are more varied than those used for input, but common experience offers only the line printer, alphanumeric terminal, teletype, plotter, and graphic display. The Responsive Environment adds electronic sound synthesizers, projectors, lighted floors, and laser beams.

Programming

The Environments are not suggested as a means of programming, but as a much richer and more involving way of experiencing a running program than is usually available to the programmer. Programmers share a common need when they are debugging a program; they must be able to see what it is doing. The programmer works under the worst possible feedback conditions, anticipating every contingency ahead of time and having difficulty determining what actually happened after the fact. The relationship between the individual and the computer guided my work long before GLOWFLOW. Since I felt that operators, Input/Output clerks, and the constraints of batch processing and time sharing all distorted this relationship, I resolved to focus only on machines that

I could use alone. This approach led to the development of a teaching system that provided a more responsive environment for programmers.

In 1967, I implemented a teaching system that sought to enrich the programming environment by providing light and sound displays through which programs could be monitored as they were running. Ironically, early machines provided more feedback than contemporary ones as they were often equipped with a speaker tied to the accumulator. Programmers came to know the proper sounds for their program to make; if an unfamiliar one occurred, it often signalled the presence of a problem. This experience is still available for the programmer of a large system who is present as the job is run. Watching the console lights, tapes, disks, and other peripheral equipment yields information on what the program is doing and at what point it goes wrong.

In my teaching system, the programmer was also given more versatile ways of controlling the execution of a running program through what was then a sophisticated debugging tool that could control the speed of execution both forward and backward. After the initial instruction, this system assisted the user in the context of the current problem. Users each had a personal teaching system on tape, which kept a running record of their current knowledge. If help was needed, the teaching system was called in. Otherwise, it kept out of the way.

VIDEOTOUCH — Computing by Hand

The teaching system just described sought to enhance the users' ability to perceive and control their programs as they were running. However, rather than have us experience the computer's programming world, we could program it to create a simulated world within which we could perform our task using familiar physical actions.

We all share a common heritage of manipulating physical objects in the world, whereas the abstract reasoning skill required for computer programming is hardly universal. In addition, many of the tasks we perform with a computer are very similar to those that we previously did by hand. For instance, computerized text editing has the same intent as the writing we do with a pencil and paper. The difference is that the act of writing is a perceptual task performed with real objects that can be physically manipulated before, during, and after the writing process.

While the use of a keyboard for typing is very efficient, its use for anything but character input is less than ideal. Editing commands are invoked through awkwardly positioned special function keys or require the totally arbitrary combination of two or three keys. In particular, the movement of the cursor to a point on the screen can require a large number of separate commands, when one could more naturally point directly to the spot.

An alternative approach would permit users to employ the same gestures they would use to explain desired changes to a human secretary. The user would work on a traditional keyboard inset into a large flat work surface and facing a large projection screen. The keyboard would be used for inputting large blocks of text. Editing existing text would be done on the flat surface surrounding the keyboard. This surface would provide a uniform color background when viewed by video cameras from above. The image of the user's hands would be inserted into the picture of the text which would be rear-projected on the large screen in front of the user (Fig. 10.1).

A set of natural physical movements would have to be established that the computer could recognize and effect as manipulations of the text. Positioning the cursor by pointing is the simplest example. Another is scrolling the text in either direction by moving the index finger or thumb vertically along the right margin. To make room for the insertion of new text, the user's hands could approach the text from either side and push all the text below the point of contact down the screen. To delete a large block of text, the two hands could enter at different points and squeeze together. For local deletions, the user could rub his or her finger back and forth over the appropriate area as if erasing. To move a block of text, the user could squeeze the text as if to delete it, but

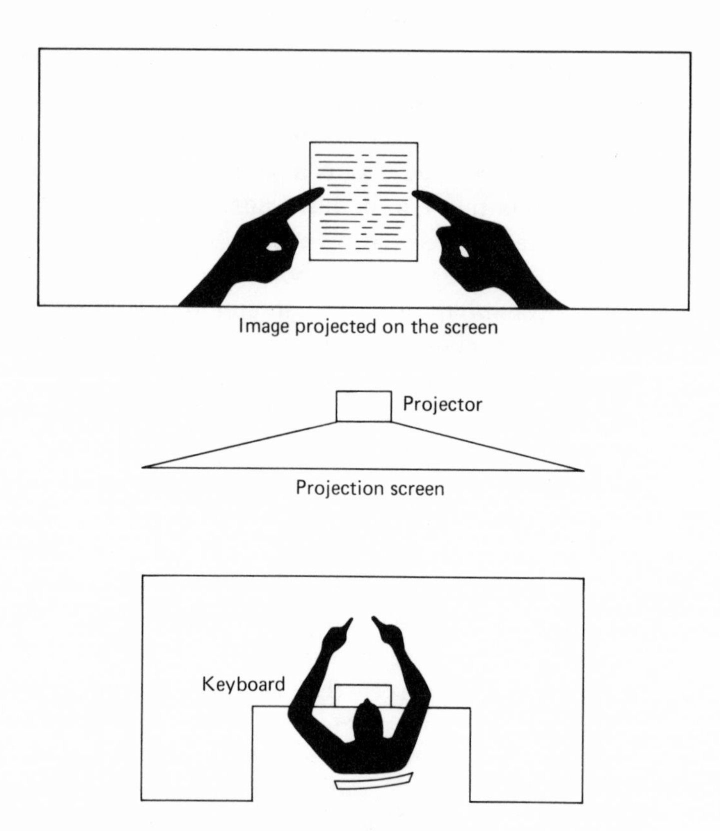

Fig. 10.1 Physical layout of workspace

stop short of bringing the hands together. Then the user could push the reduced text off to the side while finding and preparing the space to insert it. Finally, after completing the file, the user could squeeze it from the sides and then from the top to make it a small object. Then the user would open the visual equivalent of a filing drawer on the side of the screen and push the manuscript into it at the desired location.

Two-Dimensional Graphics — By Hand

The same ideas could be used with graphics. There are a number of applications where one creates a design in terms of two-dimensional graphic primitives, e.g., circuit schematics. To address this need three screens and three work areas would be necessary. One area would contain the inventory of graphic primitives, a second would be a work area where the user could manipulate a single element, while the third would contain the composite image of the complete design so far. Users would reach into the warehouse and select a graphic object, for instance, a box. They would then move it to the work area where they could take the archetypal box and make it into the specific box they wanted. They could put their hands flat on either side and slowly move them away, stretching the box horizontally. They could approach the box from two sides with their index fingers. Then by rotating the fingers around the box, they could indicate the orientation they wanted (Fig. 10.2). At this point, they

Fig. 10.2 Rotating and enlarging an object

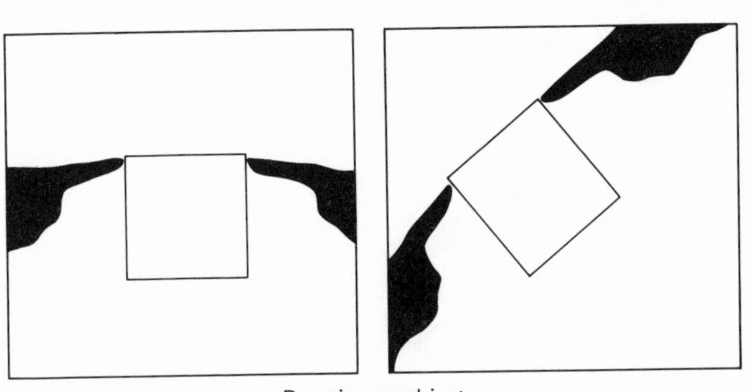

Rotating an object

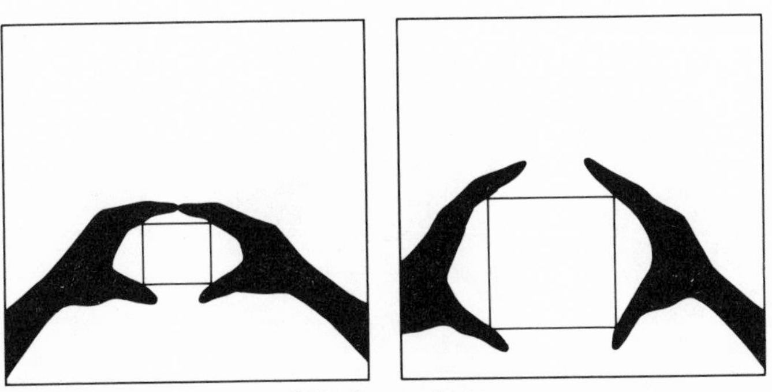

Enlarging an object

could add text fields if needed and then place the component in the design area.

In the event that the user wanted to create their own shapes, they could use an index finger as a pen and draw freehand. This first effort could be shrunk down to symbol size or expanded to allow the user to add detail to part of the shape. For more sophisticated drawing, one could use B-Spline curves.[1] These techniques use a small number of control points to define the shape of a complex curve smoothly. Each point along the curve is a weighted function of the control points. While traditional input techniques allow a person to move only one control point at a time, four points could be positioned simultaneously by using the index finger and thumb from each hand as control points (Fig. 10.3). By selecting control points along an existing curve the object

Fig. 10.3 Manipulating graphics by hand

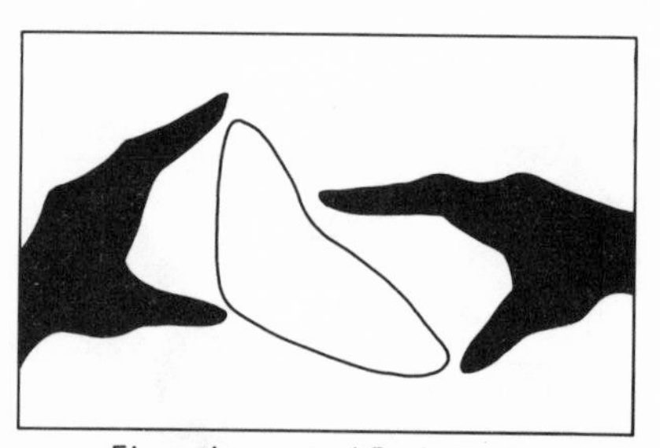

Fingertips control Bezier curve

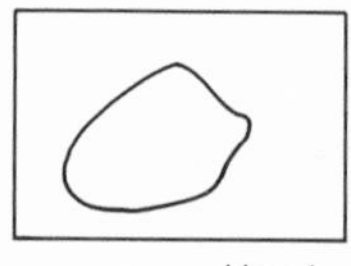
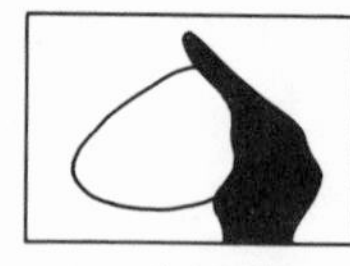
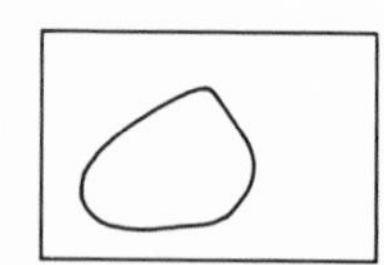

Hand used to compress boundary of graphic object

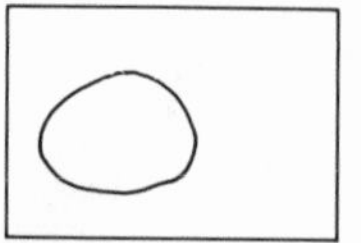
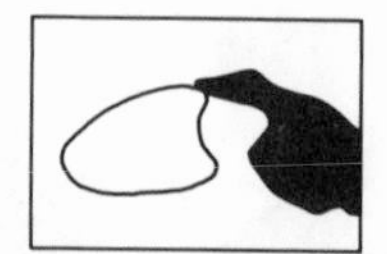
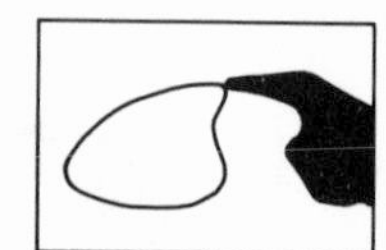

Hand used to locally stretch boundary of graphic object

could be shaped locally. Additional features would allow the object to be stretched or pinched or squeezed.

Three-Dimensional Tasks

In cases where users wished to fashion a three-dimensional object they could use their hands to sculpt it. The system would determine the precise three-dimensional position of the users' hands with respect to the graphic object. Then, as hands impinged on the space occupied by the object, the users would change the object's shape; for example, picking up graphic material and applying it to the surface of the object to build it up. Alternatively, three-dimensional B-Spline techniques could be developed that would allow users to attract or repel the control points that define the surface they touched. For fine work, they could increase the size of the object or, alternatively, reduce the size of the image of their hands and thus the scope of their effect.

The computer would have to give the user visual feedback showing their hands in relation to the object (possibly including the occlusion of the hands by the object or the shadowing of the object by the hands). The system should provide audio feedback that signals proximity and touch differently, depending on which finger or hand is involved. It would be helpful if the user wore special gloves that were color coded to assist the pattern recognition process and to provide electrical stimulation to the hand in lieu of true tactile feedback.

A beneficial byproduct of developing such systems would be an understanding of human physical behavior in the traditional programming environment. There is great interest in creating computer systems that are easy to use. One problem with such efforts is that the computer is unaware of the difficulties the user may have in using devices such as the light pen or trackball. If the computer could understand the user's physical actions, it could offer assistance when they seemed to be having trouble. Such an intelligent interface will be just one feature of the nascent trend toward focusing more computer power on the needs of the interactive user. As the designers of systems realize that happy programmers are good for sales, the standard human-machine interface will be a Responsive Environment.

Real-Time Systems

The implementation of Responsive Environments requires addressing a number of computer science problems in very general ways. Among these problems are: real-time operating systems, specialized systems, real-time simulation, real-time graphics, real-time pattern recognition and process control. Obviously, the salient characteristic of this medium is its commitment to real time — not sort-of real time, but unfailing real time. We might even call it "now-processing." While real-time systems are typically considered outside the mainstream of computer science, understanding them is crucial to the mass application of computers.

In systems research, the emphasis has always been on "general purpose" systems. This euphemism refers to the kind of system found in university computer centers. It is understandable that such systems have been well studied because they are the ones most familiar to the researchers. However, to understand operating systems in general, it is necessary to understand the various forms operating systems take when they are designed with different constraints. The operating system required for real-time response of a dedicated system in a Responsive Environment is very different from that of a time-sharing system or batch processor.

While the Responsive Environment is a specialized system, its intent is to be a flexible and therefore general specialized system. At any instant it is totally specialized. However, it must be possible to specialize it to a wide variety of interactions. Setting this goal has a tremendous impact on both the hardware architecture and systems software design. While many real-time systems exist, most of these are inflexible single-purpose systems. The ultimate goal of the systems development for Responsive Environments would be a family of hardware and software modules that can be tied together with the same facility as the logic functions we use in digital design. Because of the flexibility required for Responsive Environments, the primitive functions that suffice for their needs should cover a wide class of similar applications.

Simulation

The Responsive Environment resembles a real-time simulation system. Not only do the Environments involve all the problems of a simulator, they add the further dimension of interactive composition. Whereas a flight simulator only has to represent the kinds of interactions that a pilot might have with the controls, the approach of Environmental software is to be completely device and interaction independent and to be capable of presenting any relationship that might be found in any environment, real or imaginary.

Since the Environment is not limited to simulating real physical situations, it can be thought of as a tool for exploring the kinds of realities to which a person can adapt and therefore the kinds of systems a person can understand. Such a general approach will undoubtedly benefit simulators by attacking their problems in a larger framework.

Artificial Intelligence

Most successful Artificial Intelligence projects operate in a limited domain of knowledge about which the computer can reason or converse. The Responsive Environment is such a microworld. In it the computer must understand human behavior as it relates to graphic objects.

While most Artificial Intelligence (AI) efforts are fragmented with each researcher investigating a small aspect of intelligence and often just one step in its processing, the Environment creates an integrated entity with perception, goals, and behavior. It is a framework within which the control structures become as important as the separate intelligence functions.

Traditionally, AI programs are slow. Their only time constraints are the patience of the investigators and the tolerance of their system's other users. Contrast this situation with natural intelligence systems. All of these have evolved with the requirement that they operate in real time. The need to survive always comes first. Reflection and long-term learning have survival value but only if they are relegated to background status when an immediate threat or need arises. The leisurely intelligence of today's systems would have doomed them to extinction if they had appeared in the real world.

Human intelligence operates in a world too complex to be comprehended fully at any given moment. Given time, sense can be made of most experiences. However, there are many occasions when there is simply not enough time to process a situation thoroughly before an action is required. Therefore, a real-time system will always make a distinction between the processing needs of the moment and the longer-term processing of which it is capable.

A real-time AI system must consider the time available for a particular processing problem before choosing the best means of solving it or, indeed, before deciding whether to solve it at all. It must be aware that its higher level faculties cannot be used simultaneously with behavior in time-constrained situations. The only way such a system can behave intelligently in these circumstances is to have a prepared set of reactions that focus all of its perceptual resources on the analysis of a narrow set of expectations and a fixed set of responses. To have such a set of prepared responses, real intelligence should continually ponder its experiences after the press of real-time events has subsided. The first Artificial Intelligence discipline that will be affected by this process is pattern recognition.

Pattern Recognition

Because Responsive Environments are dynamic real-time systems, a new approach to pattern recognition is required. The traditional approach to pattern recognition has been to work with static images and to process as long as necessary to understand a scene fully. In the Environments, the parameters of the recognition process have been changed to suit the task in the following ways:

1. The subject of recognition is the human body.

2. Distinction of the human form from the background is finessed by making the background neutral.

3. The responsive system is interested in dynamic as well as static analysis of the human form. Human movements are to be understood in terms of walking, jumping, touching, and gesturing.

4. Recognition is to be accomplished in real time. This means that when a participant is moving quickly, the pattern recognition is necessarily rough. If they move more slowly, it can be more sophisticated. The purpose of the pattern recognition is to guide action. Thus, if the need to act is detected before the visual analysis is completed, the decision to act will be made on the incomplete analysis. Sometimes these decisions will be wrong, just as human decisions might be in a similar situation.

5. At any instant, recognition is not fully general. The system is interpreting its input in terms of the current interaction and on its knowledge of the moment immediately before. Its attention may be focused on a small part of the screen to see if a particular event is about to happen. It may completely miss events that occur elsewhere. Thus, the computer is not analyzing a scene in the general sense. It is looking for the significant features of its prior understanding. It then checks to see how these have changed and then interprets the significance of these changes.

6. The behavior of the human participant is to be understood in relation to the objects and the scene that are defined by computer graphics and in which the person's image has been inserted. Since the computer has defined the graphic objects it has no problem recognizing them. Therefore, it is free to focus on more structural relationships between the participant and the microworld.

It must do this in order to maintain the appropriate response of the graphic microworld to the actions of the participant. If there is an object on the screen and the participant moves their hand to it, the system must decide how the person is touching it, which must be determined by how they have moved. Is the participant hitting the object? Or just tapping it? Or trying to pick it up? If so, when should the object and hand part company? How does the system distinguish between carrying the object, dropping it, and throwing it? In addition, the system must consider the object. If it is inanimate, it is passive. If it is a bird, it may fly away. It is important for the Environment to understand the semantics of events in its microworld, for if it does not, it cannot maintain a convincing version of reality.

Model of Human Aesthetics

There is another way in which the Responsive Environment can be used to study intelligence. As an art form, it is designed to address our intelligence directly. The Maze in PSYCHIC SPACE was not arguably intelligent and yet it clearly was exhibiting the interactive intelligence of its creator. In fact, the participants were not sure if there was a human intelligence controlling it directly as there had been in METAPLAY. Therefore, the question arises, is it possible to design an artificial intelligence that will create and control experiences for the environment? Such an intellect would have to be given a context with a set of ingredients that it could manipulate and understand in semantic terms. It would have to have:

1. A concept of timing.

2. A concept of a unified experience including an understanding of:

   a. The need to introduce new concepts.

   b. What constitutes an interesting interaction.

   c. Some means of guiding the participant.

d. Some means of recognizing when its efforts were failing.

e. Strategies leading to a satisfying conclusion of the experience.

3. A way of inferring a model of the participant's current intentions and expectations from their behavior.

4. A concept of surprise and even the model of a sense of humor.

Such a system would be pursuing profound issues in Artificial Intelligence. It would be dealing with the organization of intelligence as part of personality rather than as a reasoning machine. The techniques developed in the Responsive Environment for making the computer interesting and playful might later be applied to robots. For the Responsive Environment as an art form, the future clearly lies in this direction.

Robotics

Essentially, the Responsive Environment is an immobile robot given the task of monitoring and interpreting complex and dynamic human behavior. Although we typically think of robots as mechanical monsters stumbling around on wheels or metal legs, there will actually be many more environmental robots like Hal, the robot in the movie *2001,* than mobile robots for the foreseeable future. This is because electronics technology is progressing more rapidly than mechanical technology or portable power storage.

From a control point of view, the Responsive Environment is exactly like a robot. Both have the need to assimilate sensor information and generate responses rapidly. Both must interface higher-level cognitive processing with low-level reflexes. Both may have goals that guide their behavior. The only differences are in the response modalities. The Environment lacks physical mobility and the ability to manipulate the objects of the real world directly.

Although the Responsive Environment lacks physical mobility, it has a compensatory ability to "go see." This can be accomplished by selecting information from a large number of sensors that give it alternative perceptions of the participant from different vantage points. It can also subject its raw input to differing kinds of processing.

In the same vein, although the immobile robot cannot physically manipulate people or objects, it can purposely elicit specific behaviors. The example of getting the participant to move to the starting point of the Maze demonstrates that the computer can effect changes in its environment.

In addition, as the Responsive Environment's ability to understand human movement improves, the techniques developed can be used to control a robot directly. Given the ability to perceive human hands in three dimensions, described earlier in this chapter, it would be possible to control a robot's hands. A person would operate the robot by moving their hands through the motions desired. If the robot, which could be in the same room or a thousand miles away, understood these motions, it could then be shown how to assemble a motor just as a human would be shown. The robot would remember the sequence of steps, rather than the microscopic details of the actions.

The Responsive Environment, representing the artificial intellect in the malleable stage of its infancy, is an ideal place for the scientist and the layman to get to know and participate in raising this form of intelligence to a maturity that reflects human values.

THE RESPONSIVE ENVIRONMENT AS AN EXPERIMENTAL TOOL

The Responsive Environment is essentially a flexible tool for presenting stimuli and analyzing behavior, a generalized Skinner Box that can automate even more ambitious experiments than behavioral psychologists have done in the past.[2] It will lead to much greater experimental complexity, to experiments that take advantage of the new capabilities of the computer and push them to their limit. Rather than thinking of single or small groups of reinforcers, psychologists can examine patterns or sequences or rhythms of reinforcers. They can program the computer to respond immediately to contingencies that are beyond human reflexes, automatically taking into account probabilistic schedules and past history. Longer experiments will not be limited by experimenters' endurance. They will be able to interact with the experiment as it is going on, using sequential testing techniques so no data is taken beyond that necessary to prove or disprove a certain hypothesis.

While this might not be good design for a single definitive experiment, it would be useful in a situation where the number of variables of possible interest is too large to control exhaustively. This tack would allow the experimenter to explore a large parameter space by excluding unimportant variables and focusing attention on those of greatest effect. Such a system would aid the experimenter in hypothesis formation as opposed to hypothesis verification. Once a domain of interest was identified, a more strictly controlled experiment could be run within it.

Perception and Behavior

One of the advantages of an Environmental tool is that it gives the experimenter the ability to deal with gross physical behavior instead of the more limited button pressing or pencil pushing usually involved in experiments. Quite possibly, findings based on studies that included physical movement in a natural context would conflict with some theories of behavior based on studies of sedentary subjects performing isolated tasks. Perception is not a separate activity. Perception is a necessary part of all physical behavior and often the motive behind it. We perceive, and, based on our perception, we act. In order to execute our actions, we use perception to monitor and control them. To improve our ability to perceive, we often move toward the object of perception.

Problems of pattern recognition are related to physical behavior as acting cuts down the information-handling capacity of the brain. The perceptual process in motion is different from sitting and analyzing a photograph; it is a much rougher scanning and feature-detection process. We filter out most of the information and accept only those parts of our perception that are immediately relevant.

By bringing the study of perception into the realm of active behavior, we can study the specialized physical perception of athletes, who must use only the most highly refined pattern-recognition techniques. The kinds of information they use and the interpretations they make are hard to determine at this point because we don't know exactly what they attend to. We have the same problem Leonardo da Vinci had when he tried to study the flight of birds. He misunderstood flight because he could not slow down the process and examine its smaller elements.[3] In the same way, we have never studied perception in motion because we have never had the means of controlling stimuli and recording responses with the speed and resolution the computer permits.

By outfitting an athlete with kinesthetic sensors that transmit all of the information about their body position and showing them stereo images through easily portable goggles, we could fully control the athlete's perception and capture their responses to it. Then by giving them a perceptual task within an Environment related to their own area of excellence, we have the means of studying the athlete's process of perception. Since the computer completely controls inputs and monitors outputs, it should be possible to isolate the athlete's perceptual process. This information could be used to construct a model of human perception and information-handling capacity pushed to its limits in one domain. Studying a number of such perceptual specialists might lead to new insights and generalizations.

The Environment is also useful because it is not immediately recognized as an "experiment." Since most subjects involved in psychological experiments are quite sophisticated about the interests of psychologists, there is no real possibility of studying spontaneous behavior. While behavioral scientists are quite adept at misleading subjects about the specific focus of a given experiment, there is no denying that at some level what is being studied is the self-conscious behavior of people who know that they are participating in an experiment. An experiment presented as entertainment would be more likely to evoke spontaneous behavior and because of its uniqueness would have little difficulty attracting subjects.

Of course, there is an ethical question that must be considered. Is it right to subject people to a psychological experiment without their knowledge? While we have a reflex to say no, it is interesting to consider what our fear is in the situation. The participants have willingly surrendered themselves to what is billed as an aesthetic experiment. If it is the fact that statistics are taken that is objectionable, then we must acknowledge that impersonal statistics are already kept on many aspects of our behavior: e.g., traffic patterns, purchasing behavior, and family size. In addition, if the Environment is to learn to improve its behavior in a purely aesthetic sense, it must be able to record and analyze its experience. Nevertheless, this issue must be considered carefully by any who would consider using the Environment in this way.

GLOWMOTION as a Psychological Experiment

While no formal psychological studies have yet been done, some of the effects that have been observed are interesting, particularly since the number of people involved was very large. The thousands of people who have passed through the exhibits form a far larger sample than those of most rigorous studies.

One notable effect was observed during the summer of 1972 when GLOWMOTION, an outgrowth of the phosphors used as a back-up system in METAPLAY and PSYCHIC SPACE, was taken to county fairs around Wisconsin. The exhibit consisted of a dark space, a wall painted with phosphorescent pigment, and a very high-power strobe light. A single flash of the strobe caused the wall to glow except where a person's body blocked the light. Since the flash was exceedingly brief, the shadow could be frozen in motion. As entertainment, the interaction was very successful. People jumped, did handstands, piled on top of each other, and acted out dramatic scenes.

What was fascinating was the degree to which sex and age affected the reactions. The most playful people were young teen-aged girls. Macho "studs" in their last two years of high school, on the other hand, resisted becoming involved until it was pointed out that they could do physical things like karate kicks, basketball shots, and fight scenes. Most notable, however, was the reaction of young mothers in the presence of their children. It was almost impossible to get them involved. They would stand at the sides and direct the fun for husband and kids, but neither the families nor those running the exhibit could coax these women into becoming involved. Older women were much less inhibited and an occasional grandmother would push the children aside so she could act out her own ideas.

While no statistics were taken, this was not a subtle effect that could only be discovered through sophisticated techniques. It was universal, at least among fairgoers in central Wisconsin. In thousands of families, almost no exceptions were observed. It is possible that a mother's perception of her role can place tremendous inhibitions on her behavior. It may be important for a mother not to be seen as a child.

Whether this result could be replicated reliably or with other populations is a valid question. But even if it could not, the Environment provided a means of making a very large number of observations of narrowly defined behavior from which an interesting, testable hypothesis could be formed.

Studying Feedback and Learning

The Environment can also focus on the general issue of feedback. Experiments involving the spatial or temporal displacement of feedback can be accomplished with ease, as in the Maze where the feedback representing the person's movement was suddenly rotated ninety degrees. In fact, the Environment is an incredibly versatile feedback system allowing the experimenter almost complete freedom to assign feedback to almost any action.

While the necessity of at least minimal feedback required to do a task is recognized, the possible relation of feedback to sensory deprivation has been less explored. Early experiments in sensory deprivation indicated that the human nervous system needs a certain amount of stimulation for comfort.[4] While later results qualified these findings somewhat, it is clear that people will typically avoid the total absence of stimulation. It is further recognized that certain activities in and of themselves do not provide sufficient stimulation and so lead to a sense of frustration, boredom, or anger. For this reason, Muzak is piped into department stores and assembly lines, and homemakers turn on the television while they do housework. It might be fruitful to enhance the feedback derived from a given task so that the supplementary stimulation is a consequence of the task itself, increasing workers' concentration on their task rather than distracting them. Concrete examples of how such ideas might be put into practice will be discussed in Chapter 12.

The concept of reinforcement is crucial to much of current learning theory and, like feedback, there are many ways the Environment can be used to experiment with it. The distinction between feedback and reinforcement is that feedback is the perceptual information generated by the activity, which is required to control it, whereas a reinforcer is any event that follows a behavior and makes it more likely to occur in the future.

A simple learning experiment might assume the existence of a single reinforcer that is sufficient to elicit all the behavior involved in the experiment. In working with starved rats, of course, it is relatively safe to assume that food will be reinforcing. On the other hand, in many learning situations, like schools, the animals involved are sated. The reinforcers that would be effective in this situation are difficult to predict with certainty. Furthermore, the assumption that any given reinforcer will maintain its effectiveness over a long period of time would seem to violate basic economics; the value of any given reinforcer would be expected to wane as one's experience with it or supply of it increased. While psychologists are well aware of these facts, it is often difficult to provide a variety of reinforcers, to keep trying alternatives until one works.

Therefore, a crucial problem to all learning and education is: how do you keep someone learning? The Responsive Environment, like computer-aided instruction systems, programmed texts, or indeed any method of teaching, has as an absolute minimal requirement the need to gain and maintain the participant's interest. In fact, the Environment provides a tool for studying the problem of maintaining interest either as accompanied by or isolated from the teaching of content. The grades given in courses or the simple indication given by a programmed text that an answer is correct are often not sufficient to capture interest or to maintain the learning behavior once it has started.

For instance, consider PLATO, the computer-aided instruction system developed at the University of Illinois.[5] Some years ago, while a PLATO terminal was temporarily installed by the University of Wisconsin Computing Center, a group of other graduate students and I worked out what seemed to be a difficult problem in chemistry, given our immaculate ignorance of the subject. We felt cheated at the end when merely told our answer was correct; at that point we expected nothing less than fireworks. (More recent PLATO programs do provide more novel responses.) However, fireworks, thunder, and other flamboyant responses are just part of the Environment's wildly variable repertoire.

By constantly varying the reinforcers, the Environment should be able to get participants to persist in activities that would otherwise fail to captivate them. The simplest Environment would be based on a single button that would be used to control lights and sound through the computer. The question would be how long a participant could be induced to continue pushing the button. Many strategies might be used. A series of isolated sound responses might be followed by small patterns of lights, then by rhythmic alternation between these modes with their scope becoming larger and larger until the whole space was affected by the button. Or, the button could become the only key on a strange typewriter where the letter chosen was a function of the time that had elapsed since the last time the button was struck. The participant would continue to play with the button as long as it provided relationships that took time to comprehend and, once mastered, changed in ways that continued to be of interest. To prolong its success, the computer would have to identify the behavior that signalled waning interest so it would know when to vary its approach.

This trivial Environment controlled by a single button bears a striking similarity to a piece of music. Just as each note might be thought of as a reinforcer that induces further listening, each response by the Environment should encourage further action. For this process to maintain a participant's interest over a period of time, it must involve a subtle form of teaching. If the result of every action is either known beforehand or completely unknowable, there is little chance a participant will remain involved for any length of time. On the other hand, if the reinforcers are regular enough to create expectations about the next occurrence, both the verification and the violation of these expectations can be reinforcing.

However, expectations must be taught. Each relationship must be introduced, repeated until it becomes expected, varied around the expectation, and finally violated in a way that leads to the next relationship. Once a relationship has been learned, it can be invoked later when an unknown rather than a familiar relationship is expected. Such flirtation with a person's expectations should be studied because it provides a general structure for maintaining interest through varying reinforcers. If this structure were presented in parallel with an independent development of subject matter, the interest would help motivate learning. If the student knows that the response that reinforces each correct answer will be part of a continually interesting pattern, they will be motivated to persist out of curiosity about the next reinforcer.

It is the maintenance of interest that is motivating, rather than any intrinsic value of the reinforcer itself. Consider the well-known fact that intermittent reinforcement schedules lead to more persistent learning than other schemes. What is now being suggested is that fixed or predictable schedules are inherently more boring than intermittent ones and the interest created by variable schedules makes the learning more effective. This observation may be just the simplest case of a broader principle of composing sequences of a variety of reinforcers provided at systematically patterned intervals.

Change as a Psychological Dimension

Since Alvin Toffler's book *Future Shock* was published, one of the most ubiquitous issues in the mass media has been the subject of change. While this is usually thought of as a social or political problem, it might be instructive to study it as a psychological dimension, even as an aspect of learning. Since the computer can easily establish a situation and then alter the rules that define it in a controlled, composable way, the Environment is the ideal instrument for exploring the effects of change.

Once the computer has defined a particular set of relationships it can determine which parts are stable and which are volatile. Rules can be allowed to drift slowly away from their original definitions or can change in abrupt, catastrophic ways. Since the computer provides all of the participant's input and is aware of all of their reactions, it presumably has the information needed to speculate on the state of the participant's internal model of the Environment at any given point in time. As the Environment changes, the computer can infer the changes in this model from the changes in the individual's overt behavior.

Using such a tool, we can hope to discover how people react to change and exactly why some find it so upsetting. Is it somehow related to the problem of negative transfer or is there, in fact, a general fear of learning that always accompanies new material? Berlitz and the Army Language

Schools both assume an innate resistance to learning that must be overwhelmed by a combination of fatigue and a saturation exposure to the material.

Whatever the mechanism determining their reaction, people will differ in their awareness of change, their feelings about it, and their ability to cope with it once it is recognized. It would also be of interest to know if people need change and, if so, how often they seek it. Such information would be useful in determining a manageable rate of change for each individual. If people were classified according to their differing responses to change, it would be interesting to see if these categories correlated with the more familiar psychological dimensions.

Acculturation to a Changing World

Since we can foresee that our society will continue to change, any system of education fails that does not acculturate people to live with pleasure and satisfaction in their changing world. Perhaps, through the Environment, we can invent ways of teaching people favorable attitudes toward change and mechanisms for coping with it.

Change seems to cause problems when it occurs in an area of a person's experience that they felt was settled, where their learning mechanisms were turned off. An effort to prevent such problems should be incorporated into the original learning. For example, as a child I was taught that you could not subtract big numbers from little numbers. A few years later, I was told that you could. I was annoyed, not just because I didn't believe my teachers the first time, but because what had been presented as fixed and immutable was shown to be only expedient and provisional. Similarly, we are taught grammar as a set of correct rules and are never told that these rules are part of a primitive theory of language which computer scientists have found woefully inadequate. The dictionary is also presented as the ultimate authority on words, when in fact it is merely a single opinion that always lags behind actual usage and preserves obsolete usage indefinitely. It seems that an opportunity exists throughout education to be more modest and to explain that all knowledge is transitory and subject to revision.

Kant's *Critique of Pure Reason* begins, "There are only two things which are constant, space and time." The fact that such a seemingly safe and innocent statement could turn out to be so wrong, should have had an important effect on our philosophy of education. We must teach people to keep testing what they have learned to see if it still obtains and to unlearn and learn anew whenever a change is detected.

It should be possible to design games and other learning situations where the rules are in a constant state of flux and the most perceptive, adaptable person wins. Such games would equip people with strategies for learning dynamic rules. They would have to develop working hypotheses that they would be prepared to throw away as soon as they started to fail. Such an Environment might move slowly from one organizing principle to another, even occasionally lapsing into total chaos. However, the alert and opportunistic player would have a decided advantage over one who continued to cling to concepts that were losing their efficacy. There would also be a premium on being able to recognize certain kinds of change and anticipate what the new laws would be as they were coming into effect. Such games would help prepare people to deal with the long-range fact of change by undergoing an analogous short-term experience.

From such studies we might also learn ways of introducing changes that would alleviate the suffering they cause. Changes might be signalled as they are about to occur. Or the participant might be given a choice of changes and control over when they occur. A theoretical basis may be found for structuring Environments so the whole seems stable even while the parts are in a state of flux. The human body is an analogous, well-defined structure whose actual substance is constantly being exchanged for new material, whose pattern remains fixed even as all its constituents change. If such a way of structuring our environment or our understanding of it were found, we would be able to reconcile the changes we need for progress and excitement with the stability we need for sanity.

Kinesthesiology

The VIDEOPLACE system would be useful in the study of human movement. In this field, movements are typically analyzed by taking films and measuring the displacements of limbs from frame to frame. The outline sensor used in VIDEOPLACE would seem to provide an ideal instrument for gathering data in real time. There are several problems however. First, standard video does not have the resolution of film, so precise measurements of force and acceleration are not possible. Second, the slow sample rate of the video system means that some features of rapid movements may be missed completely as they would be in standard film. The solution would be specially designed video systems that scan much more rapidly than current systems.

Even with these disclaimers it is likely that the ability to acquire very large amounts of data in real time may override the lack of precision — particularly since the computer might provide visual or auditory displays of crucial measurements as the movements were taking place. The researcher could run a large number of experiments subject to only rough analysis and then, based on this experience, decide which tests should be repeated using very high-speed film and tedious manual measurements. Alternatively, the analysis of the film could be automated based on the software developed for the Environments.

Applications in Psychotherapy

We have found that people alone, in the darkened space of an Environment, often become very playful and flamboyant — far more so than they are in almost any other situation. This could have important implications for the psychotherapist since the patient has a sense of anonymity that may permit behavior that might otherwise be impossible to elicit. In the dark the patient is protected from a negative self-image and from a fear of other people. Darkness is also a form of sensory deprivation that might prevent a person from withdrawing since, in a sense, there is nothing to withdraw from. If they are to receive any stimulation at all, it must be from acting within the Environment. Once the person acts, they can be reinforced for continuing to act.

In the event that the subject refuses to act, the Environment can focus on motions so small as to be unavoidable. It can respond to these, gradually encouraging larger bahavior and perhaps ultimately leading the patient into extreme or cathartic action. This kind of release might be beneficial.

Some mental patients are considered disturbed because they do not trust people. It is the author's impression that in certain situations therapists essentially program themselves to become mechanical and predictable, to provide a structure that patients can accept and which then can slowly be expanded beyond that original contract. However, since relationships with people are the problem, it is possible that it would be easier to get a patient to trust a mechanical Environment and completely mechanized therapy. The Eliza program at MIT was originally presented as a tongue-in-cheek offering in automated therapy, but has since been taken more seriously.[6] Its creator, Joseph Weizenbaum, professed horror at the idea that human psyches would actually be entrusted to computers; however, I think his horror was misplaced.

A few years ago a family succeeded in the dramatic cure of their autistic child.[7] The therapy they developed required ten thousand hours of effort on one child. That such a therapy is out of the question using professional staff is obvious. Furthermore, the heroic investment of that family is unlikely to be duplicated often. Such efforts are rare in any human endeavor. Therefore, it may be that the only way to accomplish certain kinds of therapy is to take advantage of the cheap patience and consistency of the computer. If it were to work, the positive ends would certainly justify the means that only seem threatening if we automatically assume that technology must be dehumanizing.

Assume that an Environment was successful in involving the patient. Once the ability to act and trust was established within the Environment, it would be possible to slowly phase in some elements of change, to generalize this confidence. As time went by, human images and, finally, human beings might be introduced. At this point, the patient could venture from the responsive womb, returning to it as often as needed. Perhaps the Environment would do things to force the patient out periodically. The possibility also exists that we might, if economics allowed, permit people to avoid others in this way. It may be realistic to try to define an Environment that adapts to the patient rather than requiring everybody to adapt to the real world as we understand it.

Diagnosis

As suggested earlier, the Environment can also be used to diagnose and classify behavioral disorders. The Environment presents a complex situation to which a participant must react. Each person will discover feedback relationships that please and others that offend. Different individuals will have very different tastes. Experiences could be designed that offer subjects unconcious choices between different feedback relations. For instance, a person may avoid areas with overload levels of feedback and favor serene relations. Later, the same person may be bored and seek stimulation. In another dimension, people may seek areas where they have positive control of events in preference to those where the Environment reacts to them, but not in ways they can control. The sequence of choices they make would reveal much about their relationship to the world. Whether different strategies for coping with the Environment would correlate with standard diagnoses is a matter for study. It is possible that the Environment would suggest new classifications or refinements in existing ones. It seems reasonable that the aesthetics of feedback would be of interest from a psychiatric point of view. Arnold Ludwig, of the University of Kentucky, has done some work in this vein.[8]

APPLICATIONS IN EDUCATION

Responsive Environments can have an educating effect both on their creators and on their technologically uninitiated participants; both groups are given an intimate exposure to the essence of current technology that is otherwise unavailable. In addition to the aesthetic experience, the Environments can be used as a vehicle for presenting specific conceptual and metaconceptual content.

The experience itself speaks most effectively to those who are out of school and therefore have little access to assistance when they seek to comprehend new technology. While few of these people would have wanted to become scientists, they have a right to learn the current and theoretical limits of science, its short- and long-range implications and a basic understanding of its language and notation.

It would be quite reasonable to design special experiences that would demonstrate the operation of the computer visually, for by allowing participants to interact one-to-one with a program, the Environment is bringing them face to face with one of the intellectual inventions of our time.

The "program" should be explicated for much the same reasons that the Renaissance Church felt the need to have painters illustrate the Bible stories for an illiterate public. Even as it allays irrational fears and presents a certain essence concretely, the Environment should convey with a child's sense of wonder the newness of our technological world. None of us understands the full implications of what we have already created, and when we are confronted by the experiences in the Environment, we have very little knowledge to draw on. We have to explore this new artificial reality just as the cave dwellers had to explore their natural environment — because it is our nature to explore.

Educational Projects

Computer science education has long since ceased to be the exclusive interest of the computer science department. Virtually all university science and engineering departments have computers. In the near future, computing power will be routinely available in the humanities and creative arts as well. However, in spite of the proliferation of computers, it is still possible for a student to receive an advanced education that does not mention them. Such an omission would be understandable if the computer were strictly a technical tool limited to specialized use. However, the computer is a culture-defining instrument that will shortly permeate every aspect of American life. Given its encroaching significance, it is crucial that computer scientists consider the problem of reaching nontechnical as well as technical students.

One means of introducing students from diverse backgrounds to the computer is through the production of exhibits like METAPLAY and PSYCHIC SPACE, which were the results of the efforts of students in a project course I taught at the University of Wisconsin. The course taught how to use the technology involved, discussed its possible aesthetics and introduced students from various disciplines to the process of organizing a large technical project. It gave the engineering students a chance to participate in the creation of art and to gain practical experience with up-to-date techniques including machine perception, graphics, data transmission, video projection, sound synthesis and distributed processing, all very important aspects of the current industrial scene.

This practical experience produces a new kind of engineer. In the past when technology was expensive and computers cost millions of dollars, the people who used them had to be highly trained, for one could simply not afford to waste computer time. Since hardware was expensive, it was only owned by large institutions. Thus, to prepare for a standard role, the student acquired a standard set of credentials and tools. Today, however, the situation has changed: the computer has become so much cheaper that it costs less than the programmer. These new economics open up a whole new set of possibilities, all of which must be explored if our technology is to be fully exploited. Large institutions can only afford to do research that has large payoffs. An idea with a possible one-million-dollar payoff would be worthless to IBM. Therefore, we must motivate individuals to do important research on their own time and with their own resources and to form small companies around every possible application. In this situation, the best credentials are a unique set of talents, insights, and prior experience. Ideally, the student will have gained enough experience with the new technology to have their own ideas and make their own inventions. Thus equipped, the student will create a job rather than simply fill one.

For nontechnical students, the course provided painless exposure to technology. They were interested in the process of technology and in the working relationship between the technologist and the machine. They wanted to experience the Environments and to compose interactions. While very few of them became proficient programmers, most learned to operate existing programs and to check out existing circuits. To these students, being surrounded by electronics and computers and probing their inner workings with an oscilloscope was a demystifying and educating experience.

There was also a cultural effect for the nontechnical students. They were asked to consider their relationship to technology, not only in terms of what it is, but what it could be. They were asked to ponder which forms of interaction are aesthetically pleasing and which are not. It was also suggested that the problems with our natural environment partly stem from our unwillingness to address ourselves to the aesthetic design of our artificial one. Henceforth, we will spend more and more of our lives in an unnatural setting while Nature itself becomes a pet or a parasite living on the periphery of man's activity, adapting ecologically to the environment defined by human beings.

Physical Activity

Watch a group of children who are "sitting still" and it quickly becomes obvious that they never stop moving. Educators are well aware that the excess energy of young children is a force to be reckoned with. Ideally, this vitality is funneled into the learning experience. However, class size usually demands discipline which suppresses opportunities for all children to use their physical energy to learn, for the hyperactive child to vent energy constructively, for the passive child to become actively involved in the learning process, and for the handicapped child to develop motor skills.

Passive Versus Active Learning

Most of the child's time in school is spent as the passive recipient of information. Facts are sent in the child's direction with the expectation that some will be absorbed. Little or no opportunity is provided for the child to assume responsibility for learning to think. It is the rare student who risks peer group mockery by saying anything he is not certain is correct. The system rewards the right answer, not bumbling attempts at original thought. In fact, it is conceivable for a student to do quite well in school without ever making the connection that facts are a codification of human thought. In the Responsive Environment, however, the experience is novel and there are no known right answers. In addition, the anonymity of the dark room permits experimentation without fear of peer judgment and students have to make their own decisions and trust their own judgment — no one else is there.

Teaching Environments

The responsive technology is a new teaching tool that could revolutionize teaching, not by automating the classroom, but by revolutionizing what we teach. What the Environment could do best would be to present nonverbal abstract material through physical experience. This would be quite merciful for children who are restless and have trouble sitting still. The learning process would be in the spirit of the enriched environments that give infants in their cribs a variety of early experience through the manipulation of many kinds of objects. However, since the Environment allows the participant to move around and the educator to control the sequence or conditions of presentation, the concept of an enriched environment can be extended to work with older children and adults. Rather than manipulating objects, the participant interacts with simple or complex feedback relationships.

We often talk quite glibly about the information explosion, but what we seldom observe is that, if there is an information explosion, there is also a concept explosion. While we have many instructors who pride themselves on teaching concepts, the time has long since passed when we should have taken education up to a higher level of generality. We have so many concepts that their importance is reduced to the level of facts and details. We need concepts about concepts and those are what we should teach first. At some point, there may well be a survey course of available concept types. These metaconcepts would reflect the ways that human beings like to think, rather than actually telling anything about an objective reality. (For instance, we have conservation laws because we like conservation laws. And in any mathematical system some quantity is conserved.)

The content of this medium would not come from an explicit presentation of facts or even concepts; rather it would derive from relationships between action and response that are established during the experience. The experience would consist of discovering the relationships that were operating and then noticing the ways in which these change. By seeing relationships in transition, the student would notice the possibility for relations among relations. In this way, an Environment could be used to teach certain important kinds of attitudes toward knowledge, learning, and change. For example, just the welter of different interactions should communicate the fact that there are many ways things might be related, many possible ways of looking at them.

It should be possible to teach strategies for learning about dynamic relationships. Students would also be expected to discover general ways of structuring relationships they find within the Environment. After a period of such training even a young student would become a creature of considerable abstract, although nonverbal, sophistication, equipped with a rich conceptual framework within which to hang specific concepts and details. Thus, the computer Environments would act as preparation for more formal schooling or perhaps as an adjunct to it. The student would know from experience that learning is a continuous process with results that must be constantly tested and revised. Students would learn not to look for the eternally right answer but rather a temporary best answer.

Such a process is not "learning by doing" in the Dewey sense.[9] This well-meaning phrase seems to have generated nothing but exercises and menial make-work problems designed to consume the student's time. Here doing is learning — a much different emphasis. If experience is the best teacher, the Environment provides a source of abstract experience and may even be capable of teaching ideas that real experience cannot.

How can artificial experience be preferable to real experience? While real experience is a powerful teacher when it teaches, it is not necessarily efficient or reliable. It is quite possible for extended experience to teach very little indeed. Furthermore, it is not guaranteed that experience will teach desirable lessons or teach them correctly. Experience does however give us confidence that we can deal with experiences in the future. The advantage of the Environments is that compact experiences can be composed that convey the desired lessons more quickly and certainly than real life.

In order to make the generalities just discussed more concrete, the following narrative describes a possible scenario for teaching children the process of scientific inquiry and precise communication. They are placed in the role of scientists landing in an alien environment. Their task is to identify its scientific laws.

VIDEOPLACE: ANECDOTAL DESCRIPTION FOR SCIENCE EDUCATION

Seven-year-old Mike enters a dark room lit only by the glow of an 8′ × 10′ video projection screen against the far wall. He laughs with a mixture of trepidation and curiosity at his life-size image projected on the screen. He jumps in the air and his image jumps. He claps his hands and his screen counterpart applauds. He has made the acquaintance of his video alter ego, a functional extension of himself.

As his awareness of his surroundings grows, he notices that although he is alone in the room, his image has company on the screen. There are a number of small graphic creatures flying in random patterns, a nest on a rock and a cage resting on the ground (Fig. 10.4). Mike watches the creatures briefly and reaches out to touch one. It swoops out of reach. He moves towards a group of three which are flying in a loose formation. Suddenly, they dart up. He pauses and notices that a large number of creatures seem to be circling the cage. He walks towards it. When he is about one foot away, they scatter in flight. He continues walking and his foot accidentally bumps the cage — which moves slightly. He intentionally kicks it and it moves again. He puts his hand out as if he were grasping its handle and raises his arm, lifting it off the floor. When a signal indicates that his time in the VIDEOPLACE is up for the day, Mike is excited about his discoveries and wants to tell someone about it. He looks for his friend Greg.

When Kristi enters the environment she is attracted to a small round creature and wants to play with it. She puts her hand out, providing a potential perch, but it only flies higher. Undaunted, she tries the same approach with another creature and another, but they fail to respond. Tiring of this, she does a couple of cartwheels across the floor. When she

Fig. 10.4 Young scientist explores artificial reality

lands in a vertical position, she notes that her legs have disappeared behind the rock. She crouches and her whole image is occluded by the graphic rock. Intrigued by this effect, it takes her a moment to realize that the small round creatures are now flying closer to her. Her hand darts out, a "boing" sounds as she touches one, and it splits in half to form two identical creatures. She stands and the creatures fly away. She is about to crouch behind the rock again when she hears the signal indicating her time is up.

In his initial exploration of this segment in the VIDEOPLACE, Mike noticed certain things about the Environment. He moved around the room and observed that he could control his video image. He saw creatures flying around the screen which motivated him to try to catch them. He was unsuccessful, but he did discover that the cage could be moved. None of his experience existed on a verbal level until Mike tried to tell Greg about what had happened. "I saw myself on a big TV! There were lots of little flying things and I tried to catch them. And I picked up the cage by myself." Greg has only a vague idea of what his friend is talking about, but he knows from Mike's excitement that he wants to find out what it's all about.

Kristi's experience with the same programmed sequence was somewhat different. She tells Mike "You can hide behind a big rock and when you touch one of the little monsters you get two." Both children are puzzled by the other's information. Mike wants a chance to hide behind the rock and try to catch the creatures again and Kristi wants to play with the cage.

Greg enters the VIDEOPLACE and stands still to take in the scene. He notices the objects. He also sees two kinds of creatures: small round ones and long thin ones. He turns his head to watch the creatures circling the cage and out of the corner of his eye, thinks he sees his screen image move, although he is standing still. He faces the screen again and sits on the floor with the intention of holding his breath and counting to 100 to see if his image will move without him. As he counts, he notices that the creatures are flying in patterns that include lower swoops than he had initially observed. As he counts 87, the small nest tumbles off the rock and lands on the floor. Forgetting his goal, he scrambles for it on his hands and knees picking it up in a sweeping gesture which brings it in the path of a long thin creature. He'd heard Kristi talking to Mike and thinks it may split in two. It doesn't. Instead, it alights on the nest and rests.

Mrs. Burger, the children's teacher, took the minimum six credits of science her college required for a BA degree. Although she is an intelligent woman, the only thing she is certain she learned in those courses is that she was not good in science. Her sense of inadequacy in the field has caused her to avoid teaching science. When the prospect of using VIDEOPLACE for science learning in her classroom was broached, she was reluctant to try it. However, one factor intrigued her. The literature describing the Environment stressed that it could be used effectively without an extensive science background. The teacher could preview each segment and make her own investigations, observations, and hypotheses about what was going on in the sequence. Since the programmed segments were essentially designed to teach the process of inquiry and develop a sense of confidence in one's ability to think about a given situation, all the necessary information was contained in the segment. After the children had all interacted with a given segment and had discussed it among themselves, she would listen to their thoughts and conclusions and share some of her own ideas. Reassured that she would receive help with the project if she wanted it and that supplementary literature suggesting possible uses of the segments was available, she was persuaded to try it.

A wide range of ideas and conclusions emerge from this segment. On a subsequent visit to VIDEOPLACE, Kristi lies on the floor and kicks her feet in the air hitting several creatures which disappear with a "ping." Mike tries to trap some of the cage-circling creatures but is unable to. However, Greg is able to catch a small round one. The teacher helps the children to verbalize their experiences and to record what has happened.

"How many creatures are there?"

"Fifteen." "Ten." "Twelve."

"How many kinds of creatures are there?"

"Ten." "Those that go 'ping' and those that go 'boing.'"

"Disappears and splitters." "Cage creatures and small round ones."

Gradually the teacher can help the children to see the need for observing carefully, which in some cases might include counting creatures and recording information to facilitate communication.

All of the children have had a unique experience that captured their interest. Each child was free to explore the Environment in his or her own way. Each could see that their actions were vital to what occurred in the VIDEOPLACE. Understanding how the video image related to physical movement, deciding what one should do in the VIDEOPLACE and what the relationship between human and creature, human and object, object and creature, and finally between human and creature and object was, were determined initially by each child's own perceptions and gradually by discussion with peers both individually and in a group. Focus and activity within the Environment was allowed to vary with the child; however, there was sufficient overlap of experience to promote communication and to spark interest in trying another child's approach on a second or third visit.

The children have been exposed to an intrinsically exciting experience. They were given sufficient freedom to indulge their curiosity and to bolster confidence in their ability to determine what is important, how things work and their own potential for effecting what happens in the Environment. A unique feature of the segments is that they can be programmed to produce certain responses only in reaction to running, jumping, or cartwheels — actions adults typically do not perform. This would render the child superior to the adult in the ability to explore the possibilities and dimensions of the VIDEOPLACE.

Class discussion would organize the body of information the children have developed. Questioning might reveal that there are two kinds of creatures: long thin ones and small round ones and that splitting, disappearing, making a certain sound on contact, and being attracted to cages are characteristics of one or the other type of creature. An experiment might be devised to test the theory offered by one child that the creatures are able to "see" the children. If they are, is it necessary to hide behind the rock to escape their vision; or, is their perception limited to a certain distance directly in front of them? Do they have any peripheral vision at all?

Other experiments could be set up to test Greg's idea that your image can move independently of your motions and to investigate why the nest fell when Greg sat still. Is there a limit to the amount of time the Environment will tolerate inaction before initiating its own action to draw the child into participation?

Some of the children draw erroneous conclusions. Kristi's experience on her two visits to the Environment tells her that if you touch a creature with your hand it goes "boing" and divides in two. On the other hand, if you touch it with your feet, it goes "ping" and disappears. Another child disagrees, because he touched one with his hand and it disappeared. What is the determining factor? The teacher would help the children to devise critical experiments to test conflicting hypotheses. The VIDEOPLACE is a complex Environment and its causal relationships can only be discovered through a gradual and repeated process of exploration, observation, predicting, testing and discussion.

Teaching Specific Skills

In addition to the teaching of very general concepts and attitudes, the Environment could be used to present more specific content. It could be used to simulate dangerous or difficult situations. Even as simple a skill as crossing the street might be covered more thoroughly by a computer that has the patience to completely define it. The total feedback capability of the Environment, which allows it to display physical motion as sound would be useful for teaching many physical skills from dancing to golf. A person could learn any given movement by trying to duplicate its characteristic sound sequence.

"Sesame Street" achieved considerable success with the animation of words as a means of teaching the alphabet. The authority of the "Sesame Street" characters could be exploited in an interactive version of the same ideas. Words could be animated to interact with the child (Fig. 10.5). Numbers, letters, and advanced mathematical symbols could be presented as friendly playful creatures (Fig. 10.6). A background of goodwill towards symbols might produce fewer people who are intimidated by mathematics (Fig. 10.7). Mathematical concepts like parameter could easily be experienced by letting some aspect of the child's behavior provide values for one of the parameters of a mathematical expression, whose curve is plotted as they move. If several children were present, each might become one of the fixed points on a least squares or interpolating polynomial. Concepts like sets could easily be expressed by drawing Venn diagrams on the floor and making the act of walking in each domain evoke a distinctive response, whereas footsteps on the intersection would be accompanied by the responses associated with each of the overlapping domains.

Fig. 10.5 Interactive "Sesame Street"

Fig. 10.6 Animated number interacts with child

Physical concepts like viscosity and elasticity could be represented by altering the child's image in the displayed space. Spatial concepts like the transformation of one surface into another could easily be conveyed by using movement around the room to control displayed movement on the surface of a sphere, torus, or Klein bottle.

Fig. 10.7 Anthropomorphized symbols

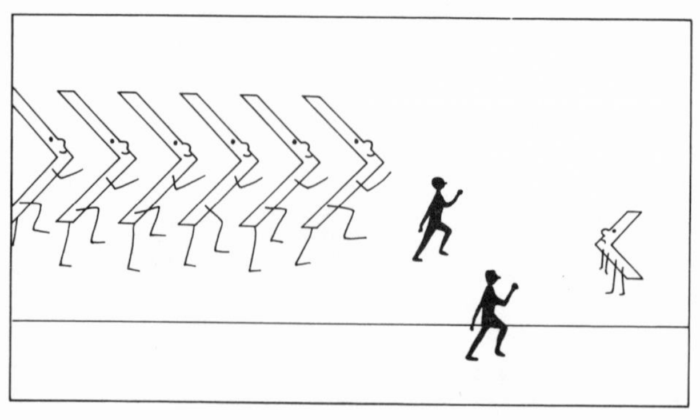

More than ...

Less than ...

The three-dimensional design tool mentioned earlier in this chapter could be adapted to instructional use. An older student could build three-dimensional molecular models by moving graphic atoms about the screen. The bonding energy of the molecule created so far would allow the student to place particular types of atoms at certain sites. They would feel the attraction or repulsion of each atom for a particular location as electrical stimuli from the gloves mentioned earlier. As the model became more complex, they could seek to change its shape to see which arrangements were stable. The student would learn essential principles by experiencing them rather than through definitions. The goal would be to develop intuition as well as intellect.

Medical students would walk through three-dimensional models of the body. They would be able to put the body to sleep, wake it up, subject it to strenuous exercise, physical trauma, and bacterial infection. All in order to see how each organ responds.

Meteorological students would be able to play with air masses in a weather model much as a child plays in a sand box. They would be able to control temperature, moisture content, elevation, and movement of air masses to see what weather would result from various interactions. The heat from the sun, the reflectivity of the earth, and the thermal drag of the oceans would all be under their control. In advanced courses, the composition of the atmosphere would be a variable as the students considered weather conditions on other planets.

Experiential Parable

Most important, but perhaps most difficult to describe, is the use of the Environment as an experiential parable. If the interaction is replete with relationships, changes in relationships, and relational ambiguity organized into a coherent whole, a sense of intellectual provocation is experienced. The result is a philosophy of events, not words. An example would be the Maze which thwarted every effort to walk it and yet used that frustration to entertain. Continuing the Maze was motivated not by the idea of reaching the goal, but by a perverse curiosity about the next frustration. Since this experience could be construed as having meaning as well as interest, if one were sufficiently compulsive about it, it is easy to envision people discussing the meaning of the Maze as earnestly as they might the meaning of a poem.

This domain will become better defined as more examples are composed. It is, perhaps, the most promising single area of exploration. The Environment might best be used to express ambiguity around a central theme. Such ambiguity is an instrument of efficient communication, for while you may not have succeeded in saying one thing clearly, you have suggested several ideas at the same time. Instruction at this level has been almost completely neglected by the institutions of our culture with the exception of the churches, which are using rather dated material. Certainly, such communication is worth attempting again and the Environment may allow us to say things we do not already know for it is fundamentally a tool for altering our own conciousness — for teaching those who created it.

PHYSICAL THERAPY

Although it may not be immediately obvious, the technology of the Responsive Environment is applicable to use by the handicapped. The VIDEOPLACE system described in Chapter 9 provides a very powerful medium for translating physical activity into participation in an artificial video reality. The participant's image is analyzed by the computer for use as a control signal and also used as an image element in the visual feedback. Both of those uses are flexible and permit modification. For instance, the input could focus on movements of the hand, finger, shoulder or face, rather than of the whole body. The output could display a graphic alter ego such as the characters used in video games, although far more interesting creatures are possible. With these modifications, a handicapped person can use the movement of any body part to communicate with the computer. These movements can be used to control the behavior of the graphic alter ego. Thus handicapped persons could participate physically in a graphic environment in which their handicap did not restrict them. They could be joined by unhandicapped people on the screen and with the help of some compensatory programming, compete in video sports.

In physical therapy, there is a related need to focus on the movement of a part of the body. Often the patient's regimen calls for endless repetition of some meaningless motion. Responsive technology could alter the nature of the task by translating it into a video context in which successive movements do not seem repetitious, because they are a means to an end within the Environment, just as walking does not seem tedious because it is seen as a means of covering distance rather than lifting the left foot and then the right foot and then the left foot. . . .

REMOTE USE

VIDEOPLACE was conceived as a telecommunication Environment. Both its educational and therapeutic uses are consistent with remote and distributed function. It would be possible for a person in one place to interact with a system or a person in another place. A possible application would arise in the mainstreaming of children with physical and emotional handicaps. These programs will create a burden for the professionals who would look after these children's specialized needs. Where the children might once have been concentrated in special schools, they are now being distributed throughout the school systems, which typically means that the support staff wastes much of its time travelling from school to school. VIDEOPLACE provides an alternative. The specialist could join the child in the VIDEOPLACE for counselling or therapy. What would be required is a camera and projector at each site and a two-way video cable system.

COSTS

How practical is a Responsive Environment like VIDEOPLACE for these applications? VIDEOPLACE appears to demand a grotesquely expensive experimental system. At the moment it does. However, it can be argued that all of the component costs are likely to go way down in the future. It is well known that the cost of semiconductor products has declined continuously over the last three decades. Therefore, it is clear that the computer processing components will not represent a prohibitive cost in the long run. Even specialized processors such as those required for pattern recognition and graphic generation should become inexpensive. In electronics manufacture, the cost of the materials is practically nil; it is development, production, and distribution that represent the overriding contributions to the total. Thus unit costs depend heavily on the volume of production. If one unit is made, the cost is X. If a million are manufactured, the cost is X divided by a thousand. Thus, if VIDEOPLACE were widely implemented, the costs of the electronics would be very low.

But what about the video equipment? In many communities the cable system is already in place and the schools already have video equipment (85 percent of the high schools in Wisconsin had video equipment in 1978). In addition, now that video disks have arrived, providing a cheaper medium than film, it is not unlikely that schools will switch to video for all mass presentations. This decision would require the purchase of video projectors for use in classrooms and assemblies.

However, a greater reason for optimism exists. What may not be apparent to most people is that video technology is subject to the same shrinking costs that affected electronics. Solid state television cameras already exist. If a mass market were perceived, the price of these cameras could come down to the point where they appear as prizes at the bottom of Cracker Jack boxes. Both flat screen displays and video projectors are also amenable to solid state design. To some extent, that is a statement of faith, for no projector manufacturers appear to be pursuing that approach. In fact, basic projector design has changed little in twenty years. However, there are experimental efforts that use a small liquid crystal video screen as a light valve through which to shine a powerful light source to create a projected image. Given the amount of activity in other areas of display technology, I am confident that some such effort will produce a low-cost video projector.

The communications costs for two-way systems may also be reduced by transmitting only the participants' images, rather than the full television picture. In many cases, their images may occupy no more than 10 percent of the screen. The graphics need not be sent at all. Computers at each end will generate the same graphic images because both are responding to the same video inputs.

Thus the technical problems are all tractable. The real problem is one of vision. If manufacturers and inventors are convinced that there is a mass market for such devices, they will exist. Thus, while at first glance VIDEOPLACE looks like a pie-in-the-sky project, it does lie directly in the path of current trends.

NOTES

1. W. M. Newman & R. F. Sproull, *Principles of Interactive Computer Graphics* (McGraw-Hill, 1979).

2. B. F. Skinner, *Science and Human Behavior* (Macmillan, 1953).

3. *The Notebooks of Leonardo da Vinci*, J. P. Richter, ed. (Dover, 1970).

4. Seward Smith & Thomas Myers, "Stimulation Seeking During Sensory Deprivation," *Perceptual and Motor Skills* (No. 3, Pt. 2):1151–63.

5. D. L. Bitzer & R. L. Johnson, "PLATO: A Computer Based System Used in the Engineering of Education," *IEEE Proceedings* 59 (June 1971):960–68.

6. Joseph Weizenbaum, "Contextual Understanding by Computers," *Communications of the ACM* 10, No. 8 (August 1967).

7. "Son Rise, A Miracle of Love," NBC television movie, aired July 31, 1980.

8. Arnold Ludwig, "'Psychedelic' Effects Produced by Sensory Overload," *American Journal of Psychiatry* 128, No. 10 (1972):1294–97.

9. John Dewey, *Experience and Education* (Collier, 1963).

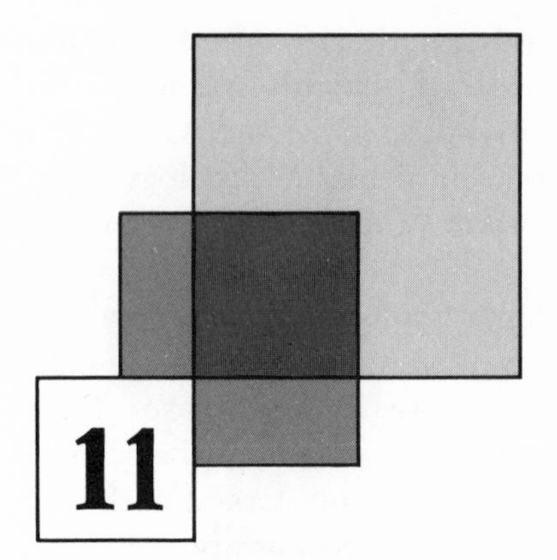

11 INFLUENCE ON THE TRADITIONAL ARTS

The traditional arts usually presuppose a passive audience that bears witness to the creativity of the artist. There is no denying that the audience may invest emotional energy in this process, but nothing in the work itself demands it. However, the possibility exists for creating dance concerts, theatrical productions, poems, and novels that require audience involvement and induce a more active aesthetic experience.

Originally, when a computer was applied to art it was used to generate a piece in a conventional format. Since then, the computer has been used to facilitate the artistic exploration of compositional possibilities. This is essentially aesthetic design automation in which the computer performs the calculations within its scope and a human provides aesthetic direction and judgment. In many of these instances, the processing systems are as interesting as the works they are used to produce.

The best way to integrate the computer and the arts is to focus on the aesthetic process rather than on producing a finished work of art. Emphasis should be on exploring the universe of alternatives that the computer can provide.

DANCE

Dancers devote much of their training to gaining physical awareness and achieving control over their bodies. One part of the body may be used to lead the other parts, or the whole body may be used to define a space. Attention may be focused on the weight of each body part, or on sustaining its momentum from one movement to the next. Such exercises are educational tools designed to give dancers new ways of thinking about the body and therefore, new ways of expressing themselves through it.

The Environment can create many such relationships by responding to footsteps, the movement of just one foot, the force of each step, posture, motion, or the distance between two dancers. Special slippers could be made to tell the computer exactly how the dancer's weight is distributed on each foot. Costumes could also be designed with colored markings on each joint and body surface, enabling a color video sensor to isolate the articulation of each part of the body (Fig. 11.1). Auditory feedback to such details of movement might prove as important to dance as the oscilloscope is to electronics. It would allow dancers to observe themselves as they were moving and thus to attain more sensitive control over each action.

The Dancer in Control

A new kind of dance could be the result of such feedback. Traditionally, a dancer moves to music, responding to the patterns established by the sound. In a Responsive Environment, the process could be reversed so that the music responds to the dancer. The dancer would play the Environment as if it were an instrument, using body movements to

Fig. 11.1 Costume to aid computer pattern recognition

create the accompanying music and to control the light displays. Feedback patterns could be established on the floor and then changed just as the audience came to expect them; or, small parts of the floor could be used for control functions that change the response relationships of subsequent actions. The audience would learn to identify such control spots and to anticipate a change in the rules as the dancer approached them. The dancer could use this anticipation as a source of tension in the piece by approaching and then avoiding the spot. The spot could suddenly cease to work; or it could turn all feedback off, signalling the end of the piece.

Posturing for purely visual reasons would become less important, because the motion of the dancer has an effective intent, to control light and sound. Far from being dominated or upstaged by technology, the dancer would control it. The Environment would amplify the effect of movement by translating it into other modalities. Such an experience could change the dancers and their philosophy of dance, for they would suddenly become very powerful figures in control of everything that happened around them. However, a new burden would be placed on the choreography, for the dance would have to sound as well as look good.

The audience could not help but realize that such a piece was less choreographed, less rehearsed, and therefore more spontaneous than traditional dance. The choreography would reflect the dancer's own plans for making use of the response relationships. If there were several dancers, each could have a unique effect on the Environment. While one was controlling the patterns of light on the floor, another could cause distortions of their projected live video images, and yet another could create the "music" (Fig. 11.2).

In the explanation above, it was assumed that the dancer knew what the Environment had to offer. Perhaps a more interesting performance would result if the dancers entered

Fig. 11.2 Dancer A distorts video image, while dancer B's footsteps trigger computer-controlled light array

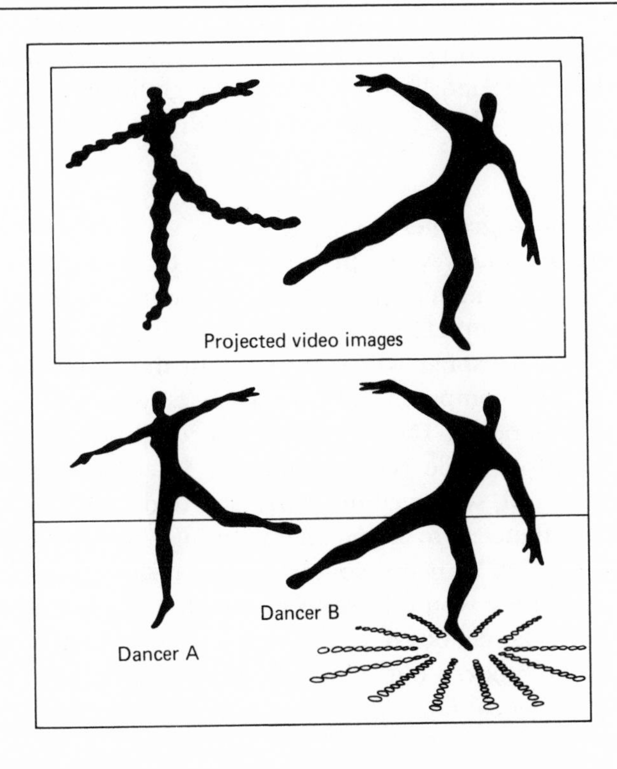

a new Environment, explored, discovered its secrets, reacted, learned more about them, gained control and finally learned to compose and express themselves using them. What began as a threatening external Environment would ultimately become an extension of the dancer. Appreciating such dances would require a completely different attitude on the part of the audience which could not expect a finished and slickly choreographed piece, but rather intelligent and graceful behavior in an unknown situation.

The Active Audience

The audience as well as the dancer could become more actively involved in the dance performance. Theater seats could be wired so the inevitable shiftings of a large group of people were sensed by the computer. These movements could be translated into sounds to which the dancer responds. Thus, the stirrings of a passive member of the audience would become an active element in the dance. Or, the audience could be given conscious control of the feedback relationships. By defining active areas of the stage, response modes, and background lighting patterns, a godlike audience could manipulate the Environment according to their whims.

The Choreographer

There are many ways in which new technology could alter the choreographer's task. With a graphic display, a dance could be completely simulated, without live dancers. Once the choreographer was satisfied with the simulated composition, the computer could generate a series of instructions for the dancers. These instructions could be translated into an auditory code representing the intended movement in great detail, which could be transmitted to earphones worn by the dancers to guide them through movements in real time. Before dancers could become radio-controlled marionettes, they would have to master the choreographer's encoding scheme, but once this was accomplished, a new piece could be blocked out relatively quickly without as much repetitive rehearsal.

The choreographer could also use responsive relationships to emphasize the movements of the dancers during their performance. If the stage had a graphic floor, patterns of light could dance around the dancer. Or, if a display that defined perspective, such as the one used in GLOWFLOW or the one projected in SPACE DANCE, were employed, the space itself could appear to dance around the dancer. Alternatively, if a number of rear screen video projections were arranged in an arc around the back of the stage, not only would the dancer appear to be surrounded by his or her own live image, but the perspective from which each of these images was taken could be changed, causing them to move around the dancer. Or, a large number of laser beams could automatically track movement causing the dancer to appear to be the center of a radiating web of light. If different colored beams were used, such that each color followed a different body part, the result would be a breathtaking undulation of swirling light.

Pseudo Dance

One intriguing possibility is the definition of a totally new kind of dance that does not involve people at all. The graphic simulations alluded to above could ultimately become so complete that they became a substitute for the dance itself. This would allow an aspiring choreographer, without access to a dance troupe, to pursue the craft. Moreover, the simulated dancers could become more than cartoon stand-ins for real people. As graphics technology improved they would become not only visually realistic, but also capable of leaps and levitations impossible for human dancers.

To carry the idea further, the computer could generate its own graphically defined entities capable of bizarre articulations. Elastic limbs, disconnected or nebulous bodies, and liquid movement would be easily attainable. The concept of dance would be generalized to include the aesthetic movement of any entities or shapes. Such shapes of the imagination could perform in a simulated graphic Environment or alternatively might frolic with the images of human participants in the VIDEOPLACE (Fig. 11.3).

Video dancer joins human dancer

Fig. 11.3 Pseudo dance

In the same vein, in the late sixties, there was a Russian proposal to physically realize a number of cybernetic animals who move in graceful ways suitable to ballet. They were to inhabit a large mechanical menagerie where they could interact with human visitors.[1]

THEATER

Responsive technology could be applied to traditional theater or the concept of theater could be expanded to include the Responsive Environment itself. In the proscenium theater the possibilities are somewhat limited but nevertheless interesting. Each actor's movements could be followed by distinctive light or sound responses that would directly enhance the characterization and become part of the character's identity. If the feedback relationship were to change, this occurrence would be noticed by the audience as a dramatic event, focussing their attention on the moment. When several actors were on the stage, the differences in the feedback that accompanied their actions would serve to accentuate the differences in their personalities or dramatic function.

The scan modulation techniques employed in VIDEOPLACE to move the participant's image about the video screen could be writ large in the theater. If just the actor's image were projected without the accompanying video frame, and this projection were deflected using small mirrors, it would be possible for the actor's image to be projected anywhere on the stage. Careful control of the projected image would allow this ghost figure to walk along walls and bound up stairs with natural movements. Or it could float about the stage, free of gravity, its position, scale, and rotation under computer control. If existing plays could not incorporate such effects, new ones could be written with them in mind.

The Audience Acts

A more interesting potential lies in the ways that responsive technology can be used to involve the audience. Past efforts to involve spectators have usually failed because the format of traditional theater includes a passive audience, a fixed script, and an isolated stage, permitting very little room for audience participation. The very existence of an identifiable cast and separate audience makes it very difficult for the onlookers to overcome their traditional passive role. As long as the physical separation is maintained, a way of involving members of the audience without them knowing is needed.

Again, as in the case of a dance performance, the seats can be wired with sensors so the squirming of each person provides signals to be detected by the computer. If these signals were translated into sound or light events, the actors could talk about these events, weaving them into the dialogue and perhaps reinforcing the person for making these sounds. Ultimately, the individual will realize who is being talked about and finally, that he or she has inadvertently become a character in the play. At this point it might be desirable to switch control to another seat.

I have used this technique when explaining PSYCHIC SPACE to a group of people who were seated on the floor within it. By focussing the computer's attention on a small part of the floor, and reinforcing movements sensed by the floor with simple sound responses, we were able to get one individual rocking back and forth without being aware of it. After a while his motions became so extreme that everyone else noticed that he was not only moving in an exaggerated way, but that it was he who was responsible for the sounds they were hearing. At this point, the computer was focused on another part of the floor and another group of people were involved in creating the sound. Soon the entire audience was crawling around trying to create sounds, trying to find the active part of the Environment. What had started as a staunchly passive audience became something quite different. In fact, the audience ceased to be an audience, thus ending the talk.

If a theater audience were seated on such a floor and disbanded by the responsive technique described above, it would be an effective way of ending a play or a scene within a play. Actors scattered throughout the audience could masquerade as part of it during the responsive period and then slowly establish a dialogue among themselves that would be recognized as a continuation of the scripted play. This part of the play, in turn, might oscillate between the structure of the script and the disorder of a spontaneous audience.

Alternatively, each member of the audience could be given explicit control of their perception of the play. If several related scenes were acted out simultaneously, the audience could not follow all the action, particularly if all the dialogue were carried on in muted tones that could only be picked up by microphones distributed around the stage. These sound signals would be available to each member of the audience through a special audio selection system that would allow listening in on what was happening to only one part of the stage at a time. Each person would see the overall gross motions that make up the play, but complete auditory information would be unavailable. Each individual would have to actively investigate what was happening, continually judge which scene was most important, and thus become involved in the production in a totally new way. This relationship places additional burdens on the playwright, the actors, and the director, for they must create a play wherein the eavesdropping audience can get a feeling for the characters and the events depicted regardless of their listening choices.

It is interesting to note that the same effect could be achieved on cable TV. A play could be broadcast on several channels, with each channel supplying a different perspective on the same sequence of events. The viewer would choose what parts to watch by switching channels.

Distributed Theater

As a result of technology, actors no longer need to be in the same physical space in order to act in the same play. It is feasible to create a geographically distributed theater, in a video space composed of video images from widely separated sources superimposed to form a single image. Each actor would enter this VIDEOPLACE through a local camera much the same as people enter a room through different doors. Images could interact visually and to some extent physically, depending on the sensing system available. The actors would be interacting solely by means of the technology. This form of theater would perhaps be most interesting if there were a mixture of real and electronic scenes. The question would always exist, "Is this event happening in video space or real space?" The audience would constantly be making a choice whether to see the live play minus some of the principal actors, or to see the whole play, via television, much as reporters sent to political conventions often end up watching them on television. Since its invention, video has been almost exclusively a vehicle for transmitting content conceived for other media such as stage or film. Since the video scene would have no real existence, distributed theater could be accomplished only through video. It would not be derivative of other art forms.

The new theater could be designed to involve people physically, since action could occur in a number of different places in one location and yet be tied together in whole or in part through VIDEOPLACE. A play could also be done in such a way that not all of the information was available through the video medium, so the audience would have to wander through the physical space in which the action was unfolding and try to learn as much as possible about what was happening, in order to understand or to affect the events.

It would also be interesting to blur the distinction between the actors and the audience further. The only definition of an actor would be someone who, willingly or not, acted in a way that was inappropriate to a person not involved in the play. Similarly, in traditional theater an actor is someone who is willing to come on stage and act as if there are not hundreds of people watching.

Spectators might unwittingly become actors in a completely unforeseen way; the computer could watch their movements around the theater space through a television camera and try to lead them to unknowingly assume a series of poses required of an actor in a certain role. The goal of this surveillance would be to collect video images of a participant in all these postures so they could be manipulated and reassembled to make the person appear to perform a completely different set of actions in a prepared video drama.

Initially, the movements of the image would be similar to the partial animation techniques associated with television cartoons like "The Flintstones." Even with such limited movements, it should be an unsettling experience to see one's own image given an independent life. With only a few frames it would be possible to recreate the brilliant mirror scene from the Marx Brothers' movie *Duck Soup*. The screen image would faithfully mirror individuals' movements until suddenly the mirror image assumed a life of its own. Such an effect is possible today, given the right aggregation of hardware.

In the future, it should be possible to interpolate from one real image to another so that the transition appears as smooth animation. Given a modest inventory of a few hundred still frames, it should be possible to make a person's image do anything imaginable.

Responsive Environments as Theater

The next step in this direction away from conventional theater is the Responsive Environment itself, which is a theater with no passive onlookers and no prepared actors. All the traditional relationships are gone; what is left is an automated happening, a theatrical situation in which the participants are responsible for their own experience.

POETRY

Kinetic Poetry — Word Dance

Poetry remains a determinedly static form, untouched by the development of kinetic media. Some attempts have been made to extend poetry into other forms. Concrete poetry, for instance, makes the shape that the poem assumes on the page part of the art form. It is odd, however, that in a day when film is probably the most accepted contemporary art form, more has not been done to animate the visual representation of poetry as is done in film credits, which often make inventive use of words in motion. It seems lamentable that no poet has seriously taken up the idea of expression through the animation of words, for such an extension of poetry seems like a very reasonable experiment.

A computer program written in 1969 at the University of Wisconsin provides an example of a simple first step. The letters of a single word were defined as a series of points, each with its own speed and direction. As long as all the points had the same speed and direction the word would move slowly across the screen. Upon reaching the edge of the screen the word was reflected point by point so that it was folded back on itself. After four such rebounds the word was restored to its original left-to-right form. After a minute or so of this process, the letters that made up the word would start to drift away from the path that the word itself had been taking. Finally, all of the points on the screen started going in completely different directions. This explosion of letters was not at all violent, but more like fluff blowing off a dandelion in slow motion. The rebounds from the edges were also a very sensitive motion. Such slow movements and transitions are more sensual than the rapid activity in most computer films.

This program suggests a whole new attack for poetry. Essentially, poetry could become a form of dance. The words and letters would constantly be in a state of flux, moving around the screen, juxtaposing with other words, transforming themselves into new words, picking up new letters and disbanding — in ways limited only by the imagination of the programmer poet. A sequence of such interactions would constitute a poem.

Interactive Poetry

Animated poetry has one disadvantage that is not as pronounced in the traditional form. Existing poetry is unique among written forms in that it is not necessarily experienced in a linear way. True, words are written down on the page in a fixed order, but as one reads and rereads a poem, the tendency is to jump around in it to better appreciate the relationships that are defined. Unfortunately, an animated poem, like a film, would start at the beginning and progress inexorably to the end. Therefore, it would be desirable to maintain the participation of the reader by making the poem interactive. The reader would enter into a relationship with the words, which would become entities moving about the screen, each with its own rules of behavior. These rules would be based on the aesthetic of the poet and the words themselves. The reader, who would be no longer a reader, but a poet as well, might most appropriately be represented by a word in this Environment of living words. The reader's word could both be changed and change others in intelligent ways as it interacted with the system. The intent of this experience would be to create a poetic interaction rather then to duplicate exactly the function of poetry.

This form of poetry need not be confined to a graphic display screen, but rather could become the basis for an Environmental experience. Graphic words and computer-generated word sounds could move around the walls, floor, and ceiling, interacting with the participant or a representation of the participant. By switching speakers rapidly, the origin of the sound could go completely around the room during enunciation of a single word. Or, if a real-time holographic projection system were available, the word could approach the participant as an apparently free-standing object. Words displayed on a lighted graphic floor could follow the participant or be chased by the participant. Footsteps could leave letters on the floor that assembled themselves into words that were spoken by the computer when stepped on. While these effects are trivial, it is probable that powerful poetic statements could be made once a suitable facility existed and there was time to explore its capabilities.

Allowing a word to interact physically with a participant is a symbolic statement, for the word is then no longer a vehicle for communicating meaning, but an entity behaving on its own. Given the impact of television and film, and the fact that computers are slowly acquiring the ability to speak and understand speech, the written word may one day be obviated. Thus, it seems appropriate to give it life, allow it to leave the page, interact with the person who wrote it, and leave the scene.

PROSE

The Novel

An interactive novel may seem a little far-fetched. But the idea is intriguing, and now practical. The simplest novel relates a single ordered sequence of events. The author may employ various stratagems for breaking through the rigid linear sequences of the time in which the story supposedly takes place: for example, alternating between parallel threads that contribute to the understanding of a larger span of events; or jumping around in time as Kurt Vonnegut does in *Slaughterhouse Five*. But ultimately an author has to yield to another form of time when writing a book, the reader's time. Each reader will read the book starting on the first page and continuing page by page until the end.

The writer typically starts with a provocative set of circumstances and a rich cast of characters, selects from this wealth of possibilities a set of consistent events, and reveals them in chosen order. Many interesting possibilities must therefore be lost. Sometimes the reader wonders what might have happened if a certain event had not occurred.

A novel published in book form is not very suitable for a nonlinear story. However, if one were to write a story that would never exist as a book, it would be natural to structure it much differently.

Alternative Story Paths

The computer provides just such a flexible means of dealing with text. There would be no difficulty in writing a story that was not a string of events, but a web of possible, mutually exclusive events that could be experienced through an alphanumeric display terminal. There would be a number of paths through the story, each representing alternative developments or perhaps different perspectives on the same events as in Lawrence Durrell's *Alexandrian Quartet*. The reader would not exhaustively follow all paths, but would explore only those that seemed interesting. Readers could follow a path until something happened they did not like, back up, and follow an alternative path where the disagreeable event had not occurred. The author could also provide all sorts of background material on the characters and situation that readers could delve into if they wished. Extraneous asides like Thackeray's or historical philosophizing like Tolstoy's in *War and Peace* would be optional; the reader could pursue them at the point in the story where they had occurred to the author, or continue the narration without this interruption. Such branching stories for children have recently appeared in paperback.[2]

INTERACTIVE NOVEL

TALE-SPIN

An Artificial Intelligence system called TALE-SPIN takes an even more sophisticated approach.[3] A cast of characters, complete with motives and goals, are modelled by the computer. They are placed in a set of circumstances and allowed to act and interact as they desire. A problem-solving program looks at the world of each character and decides what that character would do and why. The system then generates English text describing these computer events. While this system can clearly write a story from start to finish without interruption, I think it preferable not to produce a finished book, but to allow the reader to read a story as the computer generates it and be able to say "I didn't like what just happened; go back and make it unhappen and do something else." The author would design models of situations as well as a basic set of characters. The reader would follow along as the computer manipulated the model, cancelling events, removing unlikeable characters, and adding new ones from the author's repertoire. Unlikely interactions between usually exclusive characters such as a frogman and an astronaut could take place. If these options seem child-like, computers' current linguistic abilities are perhaps best suited to children. However, writing for children can provide an added benefit. Since the computer selects the words it uses to generate the story, it would be possible to introduce vocabulary at a controlled rate. Another possibility would be for the reader to enter the story as a special sort of character — one whose actions the computer could not control. Thus, the reader could act within the computer's world, engage the other characters in conversation, and impinge on their actions.

Adventure Games

The adventure games available on home computers are a step in this direction. Including elements of both story and interaction, they are played through a computer terminal. The player is the protagonist of an adventure that takes place in a world the computer describes. The game presents a puzzle or some other basis for a quest. At each step the computer describes the current situation in terms of what a player can see and the options for action. The player then chooses an action by issuing a simple command; for example, to move to another location, pick up an object, or request a clue. The computer then reports the consequences of the player's choice. Since the results of each action are controlled by probabilities, the game may be different each time it is played. It is easy to see how with the addition of more powerful displays and physical interaction, these games could evolve towards Responsive Environments.

PAINTING

While the computer has long been used to make line drawings, it is only in the last few years that one could truly simulate the act of painting. While at the University of Wisconsin, I developed a system that allowed the user to paint on a video screen using a data tablet. As the pen was moved about the tablet, pixels were filled in at corresponding locations on the screen. The color and size of the "brush" could be varied. This system was limited by the resolution of the screen (480 lines × 672 pixels) and by the number of bits per pixel (3). The latter limit meant that there could only be eight different colors on the screen at a time, although these eight colors could be any color the television screen could display and could be changed even after the picture had been drawn. Higher-resolution screens have become available as computer memory has become cheaper. These can be used to create very realistic images. With these displays, the artist would be provided a full palette of colors and a very flexible means of shading and highlighting areas that have already been painted. As memory becomes cheaper still, it will be possible to paint scenes that scintillate or change in subtle ways. Each point on the screen will not have a single static value. Rather, each point will move through a series of changes over a period of time. Water could shimmer, or the changing light throughout Monet's series of haystack paintings could be seen in a single work. As flat displays are developed, ocean scenes with waves breaking and spray flying would be hung on the wall in the traditional manner.

Computer as Artist

A very different approach to the graphic arts has been taken by artist Harold Cohen of the University of California at San Diego.[4] He has designed a system that contains a model of a particular visual aesthetic, which is used to control the generation of drawings. Since some of the choices are random, every picture is different. However, all are clearly the product of the same artist and the same style. What is interesting about this system is that it does not simply represent a step in the direction of aesthetic modeling. It works — without apology or qualification. The pictures bear the clear stamp of intelligence. They evoke a feeling of playfulness and whimsy, and they are definitely identifiable as works of art by contemporary standards. Is the computer the artist or is Harold Cohen? I would say the latter. Certainly, the original aesthetic model was his and his aesthetic judgment had to be satisfied before he would say the system worked. This is not to say that the pictures produced are those that Cohen himself would produce. Instead, he was constrained by his medium, in this case Artificial Intelligence technology. It is interesting to note that this can give the human artists a new form of immortality, for their work can continue to be created after they are gone.

FILM

Although film is not a responsive art form, interactive technology may change the movie-making process. An increasing number of film scenes are assembled through special effect techniques which are the film equivalents of the video techniques discussed in Chapter 7. It is quite common for an actor to be standing in an empty room interacting with a scene he or she cannot see. It would be valuable for both the actors and the director to be able to view a composite image as the scene was being shot or in an instant replay. In this way, they could better coordinate the actor's behavior with the rest of the film. Similarly, it would be useful if some of the other image elements could respond to actor's movements rather than trying to coordinate their actions with the other elements. For instance, if an actor is to pick up a boulder, created by computer graphics, and throw it, the director should be able to see at least a rough version of the composite in order to be able to tell the actor what to do.

Computer graphics will improve to the point where they can create any visual reality or fantasy in complete detail. Movies will take place in these graphic sets. The relationship between the actors and the scenery will be limited only by our imagination. Animation will be possible with what appears to be live images. Animated doubles and stunt men, created in the images of the human actors, will be capable of impossible feats and subject to spectacular deaths. The movie stars of the future may not be people at all, but visual simulacrums whose appearance is crafted to optimize the qualities their character requires.

RESPONSIVE SCULPTURE

The availability of new technology and a desire to encourage active interaction between a sculpture and its viewers has led to the creation of a subgenre: responsive sculpture. Among the more pleasing of the early sculptures is the "Searcher" by James Seawright which constantly scans its environment until it detects the presence of a person.[5] It then stops its scan and looks the person up and down, creating a very simple but satisfying relationship.

Another impressive interactive sculpture is a piece called SENSTER built by Edward Ihnatowitz.[6] It is a large animal-like sculpture with hydraulically articulated limbs and multiple acoustic sensors that give it the ability to select a single sound source among many or focus on certain sound levels or pitches. It has a sophisticated and variable behavior repertoire.

The SOFTWARE Show in the Jewish Museum in New York in 1970 contained what was claimed to be the first significant work of computer art. SEEK, conceived by a group of people from the Architecture School of MIT, consisted of a system that enabled a computer to build simple structures out of blocks within a small environment.[7] The space was inhabited by gerbils that were expected to keep knocking down the computer's structures. The computer was to constantly mend its walls in response to these agents of entropy. Unfortunately, according to Jack Burnham, the guest curator who organized the show, the piece had not been tested on the gerbils who stayed close to the boundaries of the space thus having only minimal interaction with the blocks that were in the computer.[8]

One responsive work that is totally unknown to the art community should also be mentioned. William Wilke, a former graduate student in physics at the University of Wisconsin, built small interactive sculptures in the early 1970s that appeared to be solid masses of rock crystals. People communicated with the stone by gently waving their hands near its ends. Sequences of these small gestures were sensed capacitively and understood by the stone which responded to each coded sequence with a specific pattern of lights. One such code included a whirring sound and the slow exposure of a previously concealed compartment containing batteries.

While the most important effect of the responsive technology on the arts will be the creation of interactive Environments, all of the other art forms should be touched in varying degrees. Equally important, each of the established media can suggest form and content for the interactive artist. We can hope that emphasis in all the arts will be less on the passive appreciation of a work by a sensitive few and more on evoking an active expression of such sentiment in many.

NOTES

1. *New Scientist,* ca. 1970.

2. Edward Packard, *The Cave of Time* (Bantam Books, 1979).

3. James R. Meehan, "TALE-SPIN, An Interactive Program that

Writes Stories," *5th International Joint Conference on Artificial Intelligence* (1977):91–98.

4. Harold Cohen, "What Is an Image?" *International Joint Conference for Artificial Intelligence* (1979):1028–57.

5. "Focus on Light" (Trenton: New Jersey State Museum Cultural Center, May 20–September 10, 1967).

6. Jasia Reichardt, "Art at Large," *New Scientist* (May 4, 1972): 292.

7. *Software* (New York: Jewish Museum, 1971), p. 23.

8. Jack Burnham, personal conversation, 1971.

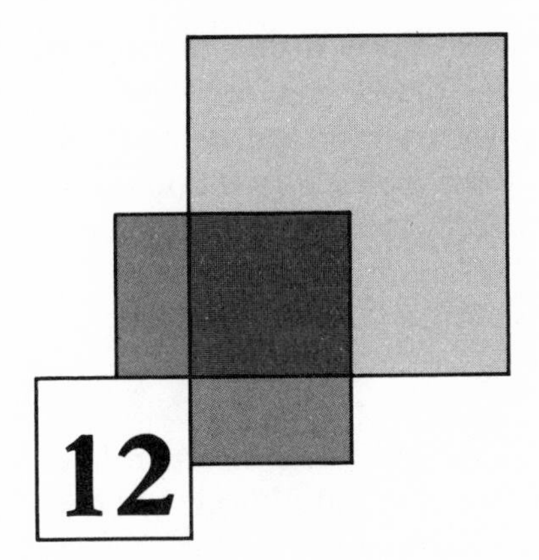

CYBERNETIC SOCIETY

During the balance of this century responsive technology will move ever closer to us, becoming the standard interface through which we gain most of our experience. It will intercede in our personal relationships and between us and our tools. The appearance in our homes of isolated devices, such as calculators, video games, microprocessors, and two-way cable TV, augurs the knitting together of a single interactive network that we will encounter through every effective device in our environment.

COMMUNICATION

While the central theme of this book is communication between human and machine, Responsive Environments also provide a general tool for communicating from one place to another. It is commonly recognized that communications and transportation are at least partially interchangeable. Recent changes in the economics of energy suggest that this substitution should be formally explored. We have already gone some distance in allowing people to act through remote control. What we must now ask is how much information must be communicated before the need to travel is eliminated.

VIDEOPLACE APPLICATIONS

The VIDEOPLACE concept could play a significant role in remote communication. People in different cities could convene for a visual conference in which each person's image was part of a large composite image (Fig. 12.1). By controlling their own video window, participants could selectively look at the other people just as they would were everyone sitting around a conference table.

In another case, a manager could visit a remote plant through video intervention. Initially, a television camera operator would walk around the site transmitting whatever the remote viewer asked to see. This camera operator would receive instructions in the form of audio signals through earphones. The signals would indicate whether the camera should hold, pan left or right, up or down, zoom, approach the subject, or move away. Observers, who might be a thou-

Fig. 12.1 Video conference

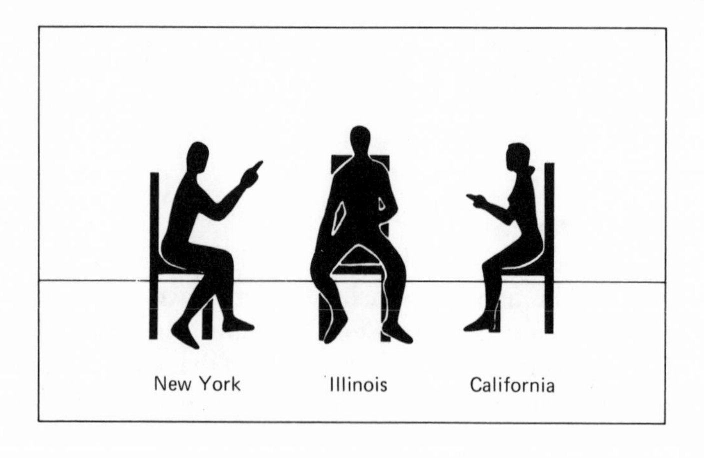

sand miles away, would interact with a video monitor, a series of simple controls, and a voice system allowing them to query those people encountered by the camera. The camera operator could eventually be replaced by a robot that might sport a video display of the teleparticipant so those at the scene could see whom they were dealing with.

Such a system would have limitations, of course. If the observer were checking the manufacture of artificial fabrics, they could sense their weight through a remote manipulator, but would be unable to appreciate how they felt. However, the transmission of tactile information is being researched.[1] The sense of actually being at the remote location could be increased if the observer wore goggles with stereo video receivers. This would create a sense of three-dimensionality and remove the distraction of the extraneous visual input at the viewer's actual location.

While discussions of telecommunication usually focus on formal communication, even business travel is usually dictated by more subjective considerations. It is often important to create a personal relationship between salesperson and customer, manager and subordinate, contractor and client. In-the-flesh meetings allow people to take each other's measure, to establish rapport, and, at some level, to become friends. It is therefore desirable for people to be able to relax together at a distance, to do things that allow them to establish a broader personal relationship which in turn facilitates professional understanding.

This need is very similar to the needs of another very large class of travellers, people visiting family and friends. VIDEOPLACE is the first communication medium that provides a way of doing things together. Friends could compete in video sports or relax over a drink at a video nightclub; astronauts could play with their children in a video playground where both parties appeared weightless. Such communication would require far greater bandwidth than the telephone. However, video cable, fiber optic and direct satellite transmission will greatly augment current communications capacity. Or, if it were imperative to conserve communication capacity, the real image would not be transmitted at all. Instead, a model of the person's appearance would be created. Then, as the person interacted, their movements would be analyzed and only the information required to update the model would be transmitted. The computer at the other end would then use its updated model to create a pseudoimage of the person that would faithfully mirror their actions and expressions.[2] It is far too soon to judge whether such a system is economically viable. All of the economics and attendant tradeoffs are in a state of flux.

State Street Mall

I proposed a variation on the theme of remote communication to the city of Madison, Wisconsin in 1975. At that time the city was planning an ambitious revitalization of the downtown area. State Street, the commercial center of the city, and the Square, which surrounds the state capitol, were to be rebuilt to form a single integrated mall that would serve as the cultural as well as commercial center of the city. One concern was that much of the population never journeyed downtown and a means was required to make that small area the focus of the community again.

The proposed solution was to make the mall acessible to everyone in town from their own homes. Madison had cable televison with dedicated but little-used public access, educational, and public-affairs channels. A large number of cameras were to be placed about the mall. By dialing the system's phone number people would be able to gain control of a camera and see what was happening on the mall on their home television. By additional dialing or touchtone commands, viewers would be able to move their vantage point from one place to another as if they were walking about the mall (Fig. 12.2). There were also to be open areas where one could call friends and speak to them through their home televisions. Mobile camera robots were mentioned as a possible later sophistication of the system. The idea was that the mall would be accessible at all times to everyone, even

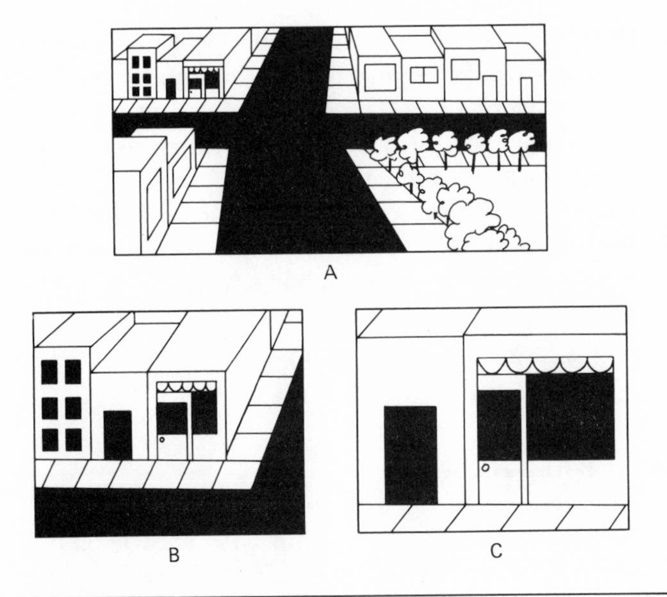

Fig. 12.2 Moving through State Street Mall by selecting different cameras

shut-ins. In this way, the spontaneous events that occurred on the mall would be part of the fabric of the community.

In the proposal I pointed out that there was a delicate privacy issue even though we were talking about a public place. I felt that while care was necessary, the system could be designed in a way that would disarm this concern. The response to the idea had been positive until I mentioned this consideration, at which point there was a reflex reaction on the part of the younger staff. They felt that even if the problem could be solved, the fact that it could be conceived made the project unthinkable. This incident illustrates my belief that for the next decade there will be more sins of omission than commission related to the use of new technology.

A modification of this proposal would be a recreation area through which people could communicate with hospital patients and elderly or ill shut-ins. In this case, anyone entering the area would understand and accept the fact that their privacy was compromised. Speakers, microphones, and cameras would be distributed throughout the park to facilitate conversations. The patient's or shut-in's symbolic ability to move, by selecting different cameras, and to make his or her location known by speaking through different speakers, would even allow participation in a child's game such as hide and seek. Such an electronically enchanted place might provide a very important sense of contact for all involved.

WORK AND WELL-BEING

Interaction and the Structure of Knowledge

In the future our relationship with knowledge will change. More information will be gained through reading display terminals and less through hard copy. We may each have our own soft copies of what we read, complete with marginalia, underlinings, and verbal notations as well as individual information systems structured to fit personal needs.

These systems will not only organize the information we already have, they will filter the information available through the media, selecting those movies, articles, editorials, and recipes that are likely to interest us. They will actively query the world's data bases, through systems such as VIEWDATA and TELETEXT, seeking answers to the questions we pose or those we would like posed.[3,4] We will all be executives served by electronic secretaries who anticipate our needs.

The Picture and the Thousand Words

Not only will the computer alter the acquisition of knowledge, it will revolutionize its presentation as well. Our most familiar and least-questioned cliche prevents us from thinking more deeply about the often awkward relationship between pictures and words. A writer does not really have a choice between the two ways of representing information; often both are necessary. The question is where to put them.

A picture with too many labelling and captioning words ceases to be a picture. On the other hand, when the two are separate, we must constantly shift our gaze from one to the other. When there are many pictures and many words, the problem gets more difficult because the picture and the words no longer fit on the same page and we must leaf back and forth between them. Not uncommonly, we make a choice. We look at the picture or the words, or the pictures, then the words. The problem is that the image and the text must both pass through the same visual channel.

The computer provides alternatives. It can represent and present knowledge not in rigidly formatted pages, but in a collage of images, text, and aural representation that can be assembled to make maximal use of a person's visual field and the spatial organization of their memory. Consider a complex diagram. Rather than constantly switching between picture and text, we can have a small text window through which the entire text is viewed, a piece at a time while the diagram is continuously displayed. Or, better yet, all text information can be presented audibly so the eyes are used solely to view the image. Arrows, blinking text, and highlighting can focus attention on parts of the image as the auditory information describes them. Dynamic concepts can be demonstrated through animation.

These techniques are in common use in educational films. While we think of such presentations as educational, we do not really recognize why they are so ideal. Nor, because of the large effort required to produce them, do we consider them part of our personal repertoire for communicating our ideas. With the computer, a single person could interactively develop such materials complete with voice and background music. Sound is useful because it provides a source of continuity during transitions in the visual material.

However, film is static in its own way. It is linear. It starts at the beginning and goes through to the end without pausing for questions or backtracking if something is missed. With the new technology, information can be made responsive. The reader can completely control the rate and order of presentation. Rather than creating a single linear exposition, an author can think in terms of a network of associations, compose explanations of all key concepts in isolation, and then provide transitions from each of these units of information to the others which are intuitively related. Readers can enter this web of knowledge at any point and follow what to them are the most logical transitions until they are satisfied that they know all that they want to know. The user can explore this network of related knowledge with the same lack of discipline they enjoy during an informal conversation with an expert.

As we combine text with verbal and animated graphic presentation, are we really constrained to communicate in the same black-and-white typography that has carried our intellectual freight since Gutenberg? It seems clear that the old alphabet may no longer prove adequate. We may, over a long period of time, evolve new symbol systems that employ color and position and movement in three dimensions to represent ideas. Instead of reading a book by a left-to-right, top-to-bottom scan, as we now do, we may enter its knowledge space and travel through it.[5] This tendency towards an ever richer representation of problems and proposed solutions may lead to the definition of problems of such complexity that they can only be attacked by the total physical as well as intellectual involvement of the problem solver, who may effectively have to live in the represented world. Each day would be spent exploring the problem space, learning about it, and intellectually and physically seeking a solution.

Among my current efforts is a project that deals with knowledge from a different perspective. This project is funded by the Advanced Research Projects Agency and is being done in collaboration with Richard Cullingford.[6] Superficially, this project is a tool for helping an electronic designer design logic circuits. In addition, it explains its operation to new users through a combination of text and animated graphic demonstrations of its features. Its goal is to provide a computer surrogate for the human expert, who ideally would sit next to new users and help them learn the system.

What is novel about this system is not what it does, but how. The explanations are not stored and read out to the user. The concepts that underlie the system's operation are represented in a knowledge structure that has nothing to do with words or a particular language.[7] The state of natural language processing is such that these knowledge structures can be used to generate correct English sentences describing their content. The generation process models users' knowledge and tells them only what they do not know. Therefore, its explanations of any given concept change as the user learns.

The philosophical implications are many, but the most significant is that a new form of human knowledge has been created, one that adapts to the learner. There are two separate constructs: the knowledge structure and the explanation mechanism. The knowledge is specific. The explanation mechanism is general. It embodies a strategy for turning knowledge of any kind into English and graphics according to the knowledge or ignorance of the learner. People who wish to record their knowledge on this system have a much different problem than before. No longer charged with creating a single linear explanation targeted for a particular audience, they must instead assist the computer in building the knowledge structures it will need to explain. By recording these structures they externalize not only their knowledge, but also their ability to explain. The computer will in turn convey that knowledge to the user in the form in which they want to receive it.

Our physical relationship with computer knowledge will also change. Today's terminal is an update of the typewriter. To use it, you must sit upright. This is not as convenient as reading a book, nor as flexible as writing with a pencil and paper. Both of these media are very portable and can be used in a variety of sitting or resting positions. The terminal of the future will easily duplicate and surpass the flexibility of books and paper. Calculators are becoming as thin as credit cards. A computer terminal of the same size can be anticipated. It is possible that an exact analog for writing with a pencil will be found, enabling the user to write on a flat screen resting in his lap. Display glasses, like those projected in Chapter 8, could certainly present text. With them, a person could lie in bed comfortably and read in any position.

A tiny camera in these glasses would allow the video touch editing system described in Chapter 10 to be used while sitting on the beach or under a tree. Voice input could be used to control the rate of presentation. Or, if a person were feeling energetic, voice presentation would allow reading while taking a walk; voice input would allow writing as well.

Enriching Repetitive Tasks

It is commonly observed that mass production, which has refined each task to optimal efficiency, has impoverished the worker's experience by not providing enough variation to maintain interest. In many cases a human is used for a function only because it is not yet possible for a machine to do the job cheaply. As Samuel Butler observed, "When a man competes with a slave, he becomes a slave."[8]

Responsive technology could be used to enliven such work, or even be used to provide some sense of community. Even within a well-defined movement, there are variations of articulation, speed, and pressure available to the human body that do not conflict with the task and might provide as many degrees of freedom as a piano keyboard. Feedback generated by the task could be reinforcing, although boredom might be intensified if the relationships were unchanging. On the other hand, if the responses were varied, a way might be found to make repetitive work interesting, if not for its own sake, then for the feedback that accompanies it. Each repetition might be approached with a curiosity about what its consequences would be. Throughout the day as the feedback relationships developed and unfolded, workers would have the feeling they were getting somewhere.

Personal Expression

If workers were able to define the feedback relationships, they could express themselves while they worked. Suppose the sounds responding to a worker's movements were not only available to that worker, but could also be heard by others so they would know what the first person was doing just by listening. Thus, each person could declare their individuality, even as they and their co-workers performed identical tasks. Alternatively, the workers could weave their feedback together, "jamming" on the job.

Another function that benefits by enriched feedback is typing. A typist always performs the same basic task, endlessly striking the keys. With an electric typewriter and especially word-processing terminals, it is easy to make the computer emit a different sound for each key as it is struck. We have done this for our computer keyboard and suspect that such a feedback would be a useful training device, providing a student with information about what had been typed without looking at the result. If secretaries were each allowed to define their own sounds, a typing pool might emit the murmur of highly individual voices rather than the consistent rattle of today.

Putting Labor Back into Work

Technology thus far has failed to relieve tedium, but it has solved one problem all too well. It has all but succeeded in eliminating physical labor from our lives. However, because our bodies require physical exertion for health, we face the irony of an extra hour devoted to exercise after we get home from work. Since our labor-saving devices have in this sense lengthened rather than shortened our working day, we must find ways to reintroduce physical effort into our jobs so we at least get paid while satisfying these undesired needs of our bodies. A way must be found to make conceptual and clerical tasks physically strenuous.

An example is the Kung Fu typewriter, which can be operated by punching and kicking (Fig. 12.3) The outline sensor is used as a control input to the typing program. An alternative would be to hold an inertial sensor in each hand that would detect the direction of any movements it was subjected to. The combination of the movements for the left and right hand would then be used to identify the letter to be typed. Ideally, such typing would be easily alterable so that one set of movements could be used for a while and

Fig. 12.3 Kung fu typewriter

then another. Such a system might not be as fast as the usual method, but by maintaining the health of a valuable person while allowing the performance of a useful task, its output might indeed be arguably greater.

Exercise

When one must get exercise apart from the job, there are ways the computer could be used to make it more interesting. Techniques very much like those suggested for assembly-line work could keep a person working out for a longer period of time. The computer could provide light and sound reinforcement as long as a given level of performance was maintained. One simple existing example is a stationary bicycle that is pedaled to generate the electricity needed to power a television. Weights and pulleys could provide programmable resistance and their movement could be used as a control input. The very simple act of weight lifting could be translated into a very sophisticated feedback relationship where the participants struggle to work their will against the complex and unwilling behavior of the load.

LEISURE

Responsive technology will affect our instruments and rituals of pleasure. Idle games, competitive sports, and musical instruments could all be heavily influenced by an electronic interface.

Games in the Responsive Environment

A whole family of games will be based on the Responsive Environment paradigm. Pinball machines and video games are but the precursors of games in which the input devices become sensors instead of controls and the displays become larger and more involving until players participate with their whole bodies.

VIDEOPLACE could be the basis of a video game where a tiny version of the person's image runs and jumps and flies about the screen. Children could play tag or hide and seek in the graphic Environment. Or, their bodies could become puzzles. They would have to discover and control strange relationships between their actions and the movements of their images.

Sports

Organized sports will be invented that take advantage of the new technology. If a laser beam were directed into an empty room occupied by two teams of six players each, hand-held mirrors could be used to deflect the beam towards the opponents' goal. A series of photocells around the reflector would tell the computer when the reflector moved so the beam could track it, thus allowing a player to run with the beam. While a number of deflections might be required to reach the goal, the team would continue to score until the opposition succeeded in stealing the beam. The visual effect would be pleasing, a jagged line of light darting about the room (Fig. 12.4).

VIDEOPLACE could be used as the basis for electronic sports in situations where people do not have room to move freely, such as in satellites, ships, and submarines. Movements within a confined space, or just changes in posture can be used to control action on the screen. The context or task provided by the computer may be contrived to lead the participant through a set of physical movements that constitute a workout. Performance in the Environment may be given an absolute score when one is playing alone or may be scored in competition with another person in a similar Environment in another part of the vessel.

Robots might also be used in sports. Today's automatic tennis-ball thrower will definitely evolve into a machine capable of playing the game. Its ability to reach the ball and the strength and placement of its return will all be adjustable, allowing the player to have an opponent of comparable skill. Remotely controlled robots with superhuman abilities may play against each other giving the armchair quarterback a new dimension in vicarious excitement.

Fig. 12.4 Laser sport

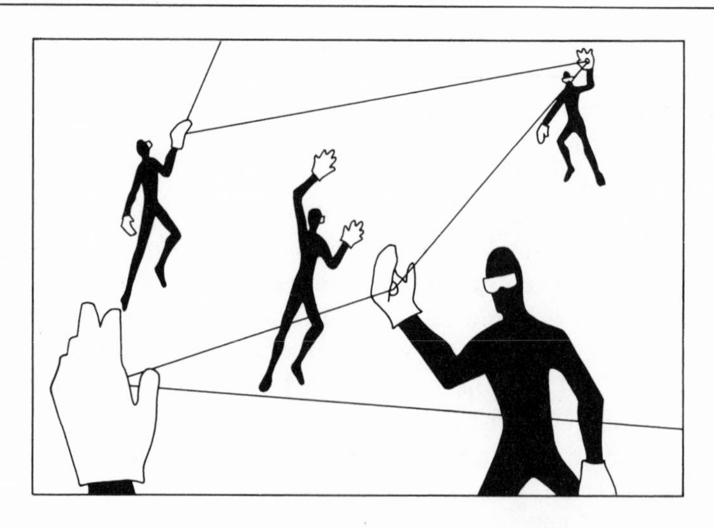

Musical Instruments

To date, the control of most musical instruments has been directly determined by the mechanical system producing the sound. Piano keys cause hammers to strike metal wires. The fingers are used to stop the holes on a flute or oboe. With electronic instruments, however, the means of control are completely independent of the means of sound generation. Thus, for the first time, the only constraint on the keyboard system is the human body. Each control system may well have its own distinctive sounds, its own range of possibilities. Control of a set of sounds through frets and strings is likely to provide quite a different idiom from control of the same sounds through a keyboard. Therefore, we can think of each kind of control as defining a class of music. Many of the sensing systems described earlier would be suitable. The video outline sensor or a series of strain gauges around the body would be versatile, although a compact keyboard like the accordion's might be more so. Certainly a rock musician would like an instrument with a distinctive sound and a unique appearance that required a dramatic set of gestures to control. The instrument would be conceived for its contribution to the visual spectacle as well as for the sounds it made.

Electron Arts Cabaret

An electron arts cabaret could add a new dimension to the concept of entertainment. To keep the emphasis on experimentation rather than on the creation of slick finished pieces, the audience would be involved in the experimental process, witnessing rehearsals rather than productions. The vehicle for encouraging such an interaction would be a circular coffee shop or cabaret with performing and display spaces radiating out from the audience seated at the core. The goal would not be to maintain interest in a constant passive entertainment, but rather to periodically dim the lights, interrupting conversation with brief experimental presentations lasting no more than fifteen minutes, a reasonable maximum for experimental forms. This format would relieve artists of the constant burden to entertain. The audience would amuse themselves much of the time, but still be available when there was something to show.

The area where the audience was seated would be wired, making them part of the environment. There would be arrays of lights in the floor, tables, and ceiling as well as keyboards, microphones, and speakers at every table. Thus, the performing artist could control the light and sound displays that moved among the audience. Or, the audience could control these displays with their keyboards, competing or cooperating with each other to achieve the desired effects. Audio or video records of their idle chatter or exaggerated gestures could be used later in the evening as part of improvised skits where their behavior was edited and comic additions and comments inserted by the artists.

RESPONSIVE ENVIRONS

The trend has already started for every effective object in our environment to be capable of perceiving its surroundings so it can perform its function as needed rather than waiting for instructions. As the economics become favorable, people will be tempted to give the appearance of life and self-expression to their most important possessions. The prime candidates are the car and the home.

The Automobile

It has long been recognized than many people view their car as an expression of themselves. This is somewhat surprising since in its current manifestation the car is a very limited means of expression, confined to a single statement made at the time of purchase which is repeated less and less strongly as the car ages.

There are times when it would be desirable to be able to express the more immediate feelings of the driver, such as embarrassment when he inadvertently cuts someone off, or annoyance when the situation is reversed. If it were coated with an electro-luminescent phosphor, the whole car could blush or turn livid. Colors could also be contingent on the direction the car was headed or the vibrations from the road. Certainly, it should also be possible for the car to be a little more tactful when reminding passengers to buckle their seatbelts.

While cars may not yet speak to each other, they are beginning to speak to their drivers. Voice maps will tell us what landmarks to expect and where to turn. If we get lost, we will describe our location and the nearest street sign and be guided back to the right path. Voice input will also replace some of the familiar driving controls. The gas pedal will disappear. We will tell the car how fast to go.[9] (Back seat drivers could cause schizophrenic cars. Who should they listen to?)

This car of the future will be fitted with radar and other sensors that detect other cars through fog and recognize rain or ice on the road. At first it will only give the driver feedback about road and traffic conditions. However, at some point, when it becomes apparent that driving has become too hazardous an activity for easily distracted humans, the car will assume responsibility for driving and navigation.

The Japanese already have a prototype system that views the road through a video camera and controls the car.[10] While such a system could not handle all driving conditions today, it is easy to foresee a highway auto pilot, akin to the cruise control that exists in many extant cars. The behavior of each car would be fit into a larger transportation system that optimizes traffic flow and the speed of each vehicle through intersections. These systems will be developed for electronic instead of human reflexes. A busy intersection will represent a split-second choreography of hairbreadth escapes performed with aplomb and orchestrated by a nearby traffic computer.

The Home

Even more elaborate will be the interaction between a house and its occupants. There are already enough control functions needed in a house to justify a central processor to coordinate them. The heating, hot water, cooling, security, laundry, cooking, intercom, lighting, and entertainment systems all contain separate control functions that could be vastly expanded with a minimum of hardware.

Almost certainly, the house will be given a voice with which to greet the family or scare intruders. When the house is empty, the computer may talk to itself in different voices, from one room to another, to create the illusion that it is occupied. Keys will be obsolete because the house will recognize it owners' voices. When the family is home, it will mediate much of the business that goes on within. The house may become a benign Big Brother, programmed by the parents to control which television programs are watched, to scold ice box forays that violate diets and to reassure children who awaken in the night. It will monitor all hazards, warning children away from a hot stove or the medicine closet. It could enable a baby sitter in an apartment building to watch many sleeping children with the help of closed-circuit television, freeing their parents to do more with their evenings.

As the artificial presence of the house becomes more convincing, it will become a member of the family. A thousand lesser electronic voices already are starting to speak to us in elevators, pedestrian crossings, self-service restaurants, and gas stations. They will speak from our watches, ovens, calculators, cameras, and indeed from everything big enough to hold a battery. Stereos will scream for help as they are stolen and textbooks will demand to be read. The utility of this feature is irrelevant. Every device is going to talk. The words will be simple at first, but the messages will grow ever more intelligent.

Shopping

One unfortunate result of our technology and perhaps more directly, of our lifestyle, has been the removal of all drama from the acts of buying and selling. If you visit the souks of Morocco, you see animation and vitality as buyer and seller compete. In this country, perhaps because we buy so much and allow so little time to do it, buying has become an assembly-line activity. Grocers know this and most supermarkets are scientifically laid out so that the customer will follow a defined path in which the order of goods is designed to maximize expenditure. Very soon our computers will make purchases for us, as is already done in industry. The refrigerator will detect the need for butter, the medicine cabinet for aspirin and the car for gas. We already are beginning to see retail transactions over two-way video cable; Sears catalogues on Video disks are on their way. Less obvious would be the use of responsive techniques in gaining and holding a pedestrian's attention at a shopping mall. An ad could gain attention with one display and then, when the computer sensed that viewer interest was flagging, switch to another. Initially, just the presence of a responsive display controlled from the sidewalk would be guaranteed to attract a crowd of onlookers.

An alternative to the Responsive Environment for sales would be the sales robot. Counterfeits of these devices already exist, promotional robots that move under remote control and allow a remote operator to converse with the people it confronts.[11] As technology improves, we can at the very least expect to see mobile vending machines stalking their prey in the shopping centers.

Electronic Boutique

The ultimate sales experience does not require true machine intelligence. Imagine an electronic boutique where a single customer enters a darkened environment. A single focussed light turns on, attracting the shopper. The light is so low that it is necessary to bend to really see what it is illuminating. In fact, to really see the object, the customer must lean on hands and knees as if peering into a pond. She can now see that it is a piece of intelligent jewelry that analyzes the stress in other people's voices and vibrates if they are not sincere.

The small spotlight on the jewelry dims. Another turns on nearby. Because of its location, recessed in the floor, it is natural for the shopper to lie down and look at it, much as a child might lie on a dock and watch the fish below. This time it is a tiny programmable light display. Although not shaped like one, it resembles a programmable calculator in that the owner can compose sequences of patterns that can be played back from memory.

As illumination by this item fades, the customer hears a sound coming from behind. It's a "Fondle," the latest electronic pet. It is round, furry, and cuddly; makes all sorts of pleasing sounds as it is caressed; needs lots of attention and care; and rolls around by shifting its center of gravity. The customer sits up to play with it for a while, and as she does, a projected image appears.

It is a simulated video store completely focused on the customer, who by voice and gestures can direct apparent movement about the graphic fantasy. Her attention is caught by a fashion show. A computer-generated model is showing off the latest styles. The model is cast in the shopper's own image, complete with the proper hair style and gait. Simple arm and voice commands will control the movements of the image. The model can even be modified to show how the garment would look if the shopper were to lose a few pounds. Since the outfit is not made until the shopper decides to buy it, it is possible to ask for modifications in the design. Buttons, belts, pleats, and hemlines can all be adjusted to taste. Once the wardrobe has been chosen, the shopper decides to consider a new hairstyle. Again, the graphic mannequin reflects the system's knowledge about her hair.

Having settled her own appearance, the shopper's attention wanders towards a display of party furniture that can be rented for the evening. Several adjunct visual spaces are created. One displays materials, another solid upholstered shapes and a third, a three-dimensional model of the customer's own dwelling. Items can be selected, covered, and placed about the room and the perspective changed to see how it would look from different angles. The model can move about the furniture to make sure that it goes with her new clothes and to see if the interactive wall and floor patterns will create the right atmosphere.

Throughout the entire experience, encouraging mood

music will be generated in response to the shopper's movements. When a purchase is made, ceremonial fanfare will erupt in congratulation. With all this attention lavished on a single customer, this store will be expensive. Few Americans will be able to shop in it. However, outside America, individual wealth is being concentrated on a scale we have never imagined, and this boutique may help repatriate some of our oil dollars.

Architecture

Both the design of future buildings and the appearance of existing ones can be dramatically affected by responsive technology, perhaps challenging our basic concepts about what a building is.

Among the most sophisticated existing interactive programs are those that aid architects and city planners. These systems allow free conceptual design while the computer keeps track of the practical constraints and alerts the user when these have been violated.[12,13] Architects no longer look on their function as static design, programming a solution to a given request for a building. Rather, they see themselves as preexperiencing a building — an important redefinition of their task.[14] There are a number of systems that allow a designer to view a space in three dimensions through stereoscopic viewers. There are also programs that allow the architect or city planner to drive through a planned or proposed design that appears not as a flat blueprint, but as an apparently three-dimensional sketch of the city. Weaknesses in the design are encountered through realistic experience during the design process. They can be easily changed since they are not yet cast in concrete. In the future, these modeled buildings could be peopled by simulated creatures programmed to search for and report the most glaring errors. It is easy to envision such cooperation between real and simulated entities becoming more complex and fruitful as the distinction between real and artificial becomes ever more arbitrary.

While the computer alone can generate a sketch of an artificial reality for a viewer or participant to experience, the video disk has an extraordinary ability to store and present real images. To appreciate its storage capability, remember that a video signal consists of 30 frames per second or 1,800 per minute or 54,000 frames on a half-hour video disk. With that number of images one could store every possible view of a real place. That is exactly what has been done with the MOVIEMAP project at MIT.[15] The computer controls which of the many frames will be shown on a projection screen controlled by the viewer. Thus the viewer can navigate a city they have never been to with complete verisimilitude. People who have used this system and then visited the locations portrayed are struck by the completeness of the experience in preparing them for their visit.

Practical construction problems have led us to create buildings with some of the most sterile interiors imaginable. The evenly lit, endless tunnels found in airports and apartment buildings provide an unbroken monotony so rigid as to be upsetting. One solution, designed by British architect Cedric Price, is a building whose walls and halls can be physically moved under computer control each night.[16]

There are also ways that the tedium of the physical space could be relieved by defining an alternative visual space within it. Abraham Rothblatt, a one-time Madison artist now living in Philadelphia, used tape applied to the floor, walls, and ceiling of an existing space to define another, often ambiguous visual space within it. Using electroluminescent tape such a display could be made dynamic, allowing the same boring hallway to appear to be different each day. The patterns could change in response to those who move through them, perhaps leading them to their room. With a powerful display system visual organisms could be defined that would effectively haunt the building's corridors, making them seem a little more lively even when empty. Each species might have its own habits either seeking or avoiding people according to its nature.

These displays might be static much of the time, but occasionally involve two strangers about to encounter each other in a mini-happening. While people are often apprehensive in such a situation, they would be brought together because for the moment they would be sharing a theatrical event that occurred solely because they were both there.

The exteriors of buildings are often monotonous as well. An exception are those whose surfaces are completely consumed with the need to advertise. In Las Vegas the buildings have been completely transformed into animated signs whose primary function is simple communication. As usual, the twentieth-century idea of communication is one-way broadcast or dissemination rather than interactive dialogue.

In 1971 I proposed a more responsive building for the Outdoor Sculpture Competition that was held that year at the University of Wisconsin. The university administration is housed in Van Hise, a monolithic twenty-one-story building with the president's office at the top. While the building is somewhat intimidating, its wide flat sides seemed to me to offer a unique surface for display. I proposed to take over the circuits in the building to allow it to acknowledge people below and to respond to their control. If the fluorescent lights in each office were controlled, the exterior of the building could be used as a giant display. A television image of a person standing outside the building would be used as input. As the person waved at the building, it would metaphorically wave back, the patterns that flowed over its surface controlled by the person's motions (Fig. 12.5). Technology would thus be used to expand the power of the individual, rather than to create a sense of impotence.

Rather than publicize the project ahead of time, I wanted to do it quietly. Then late at night, when only one or two people were walking by, the building would respond. Initial reports would be treated as less credible than UFO sightings. Then, over a period of weeks the rumors would spread until people began to hang around waiting for something to happen. The finale would be an outdoor happening in which audience and dancers would interact with the building.

Fig. 12.5 Participant controls Van Hise

Binary Icescape

When the administration failed to hand over the building, and the competition committee said they wanted something for the Union Terrace overlooking Lake Mendota, I proposed a more modest gesture, a three-mile-long light display on the frozen lake, comprised of 256 strings of Christmas lights. A participant would control the interplay of lights from the top of the boathouse, or from a plane flying overhead (Fig. 12.6).

INTIMATE TECHNOLOGY

Even man's most intimate communication might be mediated by the computer. Electronic sex could free a lover from the constraints of the human body. For example, a

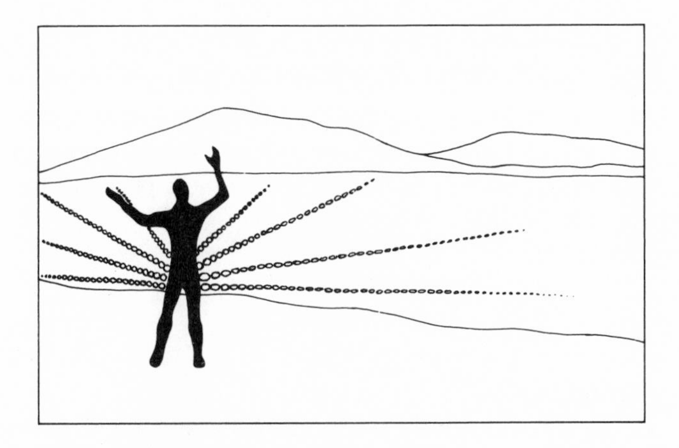

Fig. 12.6 Binary icescape

sequence of ministrations ordinarily requiring two hands could be carried out automatically, allowing the lover to lavish attention elsewhere, just as one can set up rhythms on an electronic organ. Indeed, it is just possible that sexual concerts may arouse the masses of the future. The initial reaction most people have to this thought is repugnance, but it is possible to postulate circumstances that would lead to such a development and lead many people to accept it.

It is probably easier to think in terms of technologically mediated physical intimacy outside of a sexual context. Suppose you were separated from someone you cared about, someone who needed comfort; for example, a small child who was hurt and upset. Suppose you were speaking to the child on the phone and had a device that you could squeeze, which in turn would squeeze the child's hand reassuringly. Would you use it? Most people I have talked to thought they would, if the transducers were aesthetically pleasing. (In a way, this is not surprising if you consider that between the hand and the brain all sensation is transmitted as electrical signals. All that is being proposed here is an extension of the way nature does it.) Once this intellectual step is taken, it is easy to extend the idea in small and acceptable increments. The communication can be made two-way so that each party can respond to the other, and the concept expanded from one hand to two hands. There would be a continuum of such small increments that lead to full telesexual communication. At each end the transducer could become more and more elaborate and realistic until it became a perfect model of the missing person.

Hugaphone

As the transducer became more personalized, it would be less generally useful. But it does suggest one plausible variation. We usually think of telecommunication as being a brief verbal communication. In fact, most conversation includes lengthy pauses by one or both parties. Transatlantic telephone lines take advantage of these pauses to use the cable for other transmissions. Thus, while what you say is transmitted, you never have a dedicated channel for your communication. Rather, your conversation is broken down into packets that are interleaved with the packets of other conversations on the transmission line. It would be possible to stretch the time frame to provide a new type of communication.

One of the pressures that modern life places on the family is the separation of couples because of travel. While the telephone provides contact, it is just that — a brief contact. Perhaps more important would be a less concentrated communication that would provide a continuing sense of presence even when nothing is being said.

Thus, even at night, a couple would be aware of each other's stirrings as they slept. Voice communication would always be available in the infrequent event that one of them spoke. Tactile communication, which requires rather low bandwidth, might be included to provide a physical as well as an auditory presence. The telephone network would establish a communication path between the two sites, not as a physical connection, but as routing instructions that the occasional transmissions from each location be transmitted to the other by whatever line was available at each moment.

ARTIFICIAL ENTITY

As the computer speaks and understands speech, a permanent confusion will arise about just how intelligent the computer is. It is very fashionable for researchers in Artificial Intelligence to deny that what they are creating will eventually approach true intelligence. This pessimism is partly a reaction against the embarrassing enthusiasm of the 1960s. Workers in AI point to the next problem to be solved as an argument for why the goal can never be reached.[17] Such caution is disingenuous.

Looking at Artificial Intelligence from the perspective of the last fifteen years of research, the results are not at all discouraging. Most of the initial false premises have been discarded. Purely syntactic approaches have been abandoned in favor of a head-on attack on semantics. Speech generation has become a consumer item. Language understanding, visual-pattern recognition, and chess playing all now appear to be difficult but tractable problems. The problems that remain are: structuring knowledge, learning, and problem solving. After these are solved, creativity and aesthetics will be next, but they are part of a different field — Artificial Genius.

Until recently, much of this work was done on small processors. Artificial Intelligence research has not yet felt the effect of dirt-cheap memory, multiprocessors, special hardware, or associative memory. Each of these technologies may have appreciable effect on its progress. We should start preparing ourselves now for the fact that our intelligence will no longer be unique. It is not necessary for Chess 5.0 to beat Bobby Fischer at "high noon."[18] Computers can play world-class speed chess, world-class correspondence chess, and a very respectable level of tournament chess.[19] There are millions of intelligent people who think of chess as their primary avocation but who cannot beat the best of the chess programs. Remember, chess is not just a game; it has been accepted as the highest form of intellectual competition in Western culture. Whether the computer uses the methods a person would use is irrelevant. If its methods suffice in one domain it is quite possible simple techniques applied rapidly will also work in others.

Artificial intellects will be developed to accomplish many sophisticated skills. There are systems that exist today that are better than humans at mathematical manipulations such as symbolic integration and even certain areas of medical diagnosis. It is certain that in many areas of intellectual activity, we will defer to the computer's mental skills.

While many people would say the problems of AI are insuperable, they are wrong. Consider the fact that computers operate in millionths and even billionths of a second today and will execute in trillionths of a second in the future. Our brain components, on the other hand, operate in thousandths, hundreds, and tenths of a second. Consider also that when we say computer, we mean today's device, which is based on an architecture conceived thirty years ago. In the last decade we have seen the entrance of experimental systems that depart from the one-computer, one-processor tradition. The current record holder to my knowledge is the CLIP4 system which has ten thousand very simple processors.[20] The 2″ video tape system used for storing satellite pictures while I was at the Space Science and Engineering Center at the University of Wisconsin had over a trillion bits of memory.

Thus, while our brains are comprised of billions of cells, each a simple processor, our technology is starting to catch up, especially when you consider that the design of the human brain represents a much more severe engineering problem than computer design. Our brains must be small and portable, capable of being fit into half a shoe box. They must enable every one of their myriad cells to eat and excrete. Finally, they must tolerate the tendency of large numbers of these cells to die throughout the progress of a lifetime. From a purely engineering point of view, it would seem easier to make an intelligent computer than an intelligent human being.

There are those who say the computer can never be intelligent because it does not have needs or a body or a way of fully experiencing and manipulating reality.[21] But there are people with similar handicaps. At any rate, computers are being given sensors and manipulators. They will engage the world and interact with people. Good social interaction is inevitably adaptive, and we will undoubtedly want our computers to fit in socially. As each computer learns about people, it will interpret its unique experiences individually. Thus, we can expect each to develop its own personality. Unfortunately, since it is difficult to imagine a learning program that will not go awry if faced with bad experiences, we can expect to find some computers that tend to be more aggressive, diffident, or paranoid than others.

We will certainly see systems that will be able to pass for human in a brief telephone conversation. Some people will probably demand social conventions whereby computers answering the phone identify themselves as such, just as margarine was prevented from imitating the color of butter for years. Products based on such intelligent systems will be chosen not just for their competence, but also on the basis of how well they please us. Once the intellectual skills have been reduced to computation, wit and charm will be desirable selling features. Although, if these social skills are attainable, there is reason to ponder the possibility of a super intelligence that is able to manipulate our emotions, convulsing us in laughter or reducing us to tears more easily than a human could.

Somewhat longer in coming will be mobile, physically anthropomorphic robots. Surprisingly, computer processing may not be the technology limiting their development: we have not been making the required advances in the very small portable energy sources or miniaturized mechanical controls and electronic sensors that are required.

But if we were able to duplicate the articulation of the human limbs, the seemingly hardest problem is all but solved. It might seem impossible to create a mechanical face that would express human emotion. However, the talking disembodied heads in the Haunted House in Disneyland are already disturbingly convincing.

We can undoubtedly produce a machine that satisfies our aesthetic sense. Again, there are no limits on the physical beauty of such creatures. We may be just the ugly, dumb ducklings that created them. Is there a reason to build such things? Are there reasons not to? It does not matter; they will be built. How much they are used will depend on whether people like them, as much as on whether they are effective. There is no reason to think we will choose our computer companions any differently than our human ones.

IDENTITY

The technology being described cannot be sloughed off as simple materialism that can be accepted or rejected, for it ultimately impacts our identity as well as the means by which we gain our ends. We are shaped as much by our tools as our tools are shaped by us. More and more, we are integrating ourselves and our machines. Already, medical technology allows the replacement of a host of mechanical body parts with artificial substitutes. Prototypes exist for artificial eyes and ears that bypass the original organs and stimulate the brain directly. Artificial hearts, kidneys, and pancreases are also being developed.[22]

At the moment we simply replace faulty original equipment, but brand-new organs for dispensing medication or suppressing pain have already been developed.[23] At some point it will occur to someone to include a few discretionary organs, a few extras. We can expect surgeons to insert the first calculator, radio, telephone, or electronic-scratch-pad memory in someone before the end of the century. But when will we get something even more practical, like an organ that burns up extra sugar — or a subcutaneous light display?

Faced with the loss of part of their body, most people would willingly accept a prosthetic device that was indistinguishable from it. I have asked many people if they would accept an artificial arm, leg, kidney, heart, or eye. In every case they say, yes, if the alternative is handicap or death, especially if the cosmetic problems are solved. Following this line of reasoning leads to the inevitable question, are there parts of the body that one would not allow to be replaced even if form and function were preserved? If we fully understood the structures of the brain and their specialized functions, we could probably talk about replacing each of the processing parts of that organ just as we do the rest of the body, especially if all memory and personality were preserved. Would you allow the replacement of your reticular formation or corpus collosum if the alternative were death? In fact, such an incremental approach is not necessary. Having just acceded to the idea of mechanical replacements for every other part of the body, most people sigh and acknowledge that if they had no other chance for survival, they would accept a mechanical brain.

If these observations are true, we are faced with an astonishing result. Where we might expect greater resistance to mechanical replacements of vital and intimate organs, we find that what sounds like the ultimate philosophical question, "Am I a man or a machine?" is answered by many, "I don't care, as long as I am." At the personal level, the mind-body question has been answered. Apparently, identity does not lie in "the hardware." The consensus seems to be, "As long as it's my program, I don't care which processor it runs on, if I can't tell the difference." It is interesting that the key concerns at every step are: "Would I feel the same?" and perhaps even more important, "Would others see me as the same?"

The examples given in this chapter represent only a few of the infinite number of possibilities that may be realized. Some of the examples have been deliberately trivial because our future lifestyle will no longer be characterized by a few economically compelling uses of computers, but will be permeated by a welter of ancillary applications implemented simply because they are pleasing.

NOTES

1. A. Michael Noll, "Man-Machine Tactile Communication," *Creative Computing* (July/August, 1978):52–57.

2. A. Chapanis at Johns Hopkins has made a proposal to this effect to Advanced Research Projects Agency.

3. Doug Payne, "Bye-bye Buzby, Bye-bye," *New Scientist* (May 28, 1981):539.

4. "Simple Teletext Keypad," *Popular Science* (January 1981):73.

5. Aaron Marcus, "Experimental Visible Languages," *Apollo Agonistes: The Humanities in a Computerized World* 2, (April 1979):349–57 (discusses a film related to this idea).

6. R. E. Cullingford, M. W. Krueger, M. G. Selfridge & M. A. Bienkowski, "Automated Explanations as a Component of a CAD System," to appear in *IEEE Transactions on Systems Man and Cybernetics,* special issue on human factors and user assistance in Computer Aided Design (December 1981).

7. R. Schank & R. P. Abelson, *Scripts, Plans, Goals and Understanding* (New Jersey: Erlbaum Press, 1977).

8. Samuel Butler, *Erewhon* (New York: Penguin, 1970).

9. Peter Marsh, "The Making of the Computerized Car," *New Scientist* (December 6, 1979):770–73.

10. S. Tsugawa et al, "Three Dimensional Movement Analysis of Dynamic Line Images," *IJCAI* (1979): 896–901.

11. Edward Teja, "Marrying Voice Recognition and Synthesis, Robot Pontificates on Presidential Race," *Electronic Design News* (August 5, 1980):39.

12. Clifford D. Stewart, "Integration of Interactive Graphics in the Real-time Architectural Process," *Online 72* (Middlesex: Paca-Press, September 1972):957–66.

13. P. E. Walter, "Computer Graphics Used for Architectural Design and Costing," *Computer Graphics,* ed. R. D. Parslow, R. D. Prowse & R. Elliot Green (London & New York: Plenum Press, 1969), pp. 125–33.

14. Avery Johnson, "The Impact of Computer Graphics on Architecture," *Computer Graphics in Architecture and Design,* ed. Murray Milne (New Haven: Yale, 1968) p. 57.

15. William C. Donelson, "Spatial Management of Information," *ACM-SIGGRAPH Proceedings* 12, No. 3 (1978):203–209.

16. "A Building that Moves in the Night," *New Scientist* (March 19, 1981):743.

17. H. L. Dreyfus, *What Computers Can't Do* (Harper & Row, 1972).

18. "Chess 4.7 Versus David Levy: The Computer Beats a Chess Master," *Byte* (December 1978):84.

19. "Chess Computers Start to Give Humans a Tough Game," *Electronic Engineering Times* (April 18, 1977):4.

20. M. J. B. Duff, "CLIP4: A Large Scale Integrated Circuit Array

Parallel Processor," *International Joint Conference on Pattern Recognition Proceedings* (1976):728–33.

21. H. L. Dreyfus, "Why Computers Must Have Bodies in Order to Be Intelligent," *Reviw of Metaphysics* 21, No. 1, (September 1967):13–32.

22. "An Insulin 'Pacemaker' for Diabetes Sought," *Electronic Design* (December 6, 1973):25.

23. Medtronics Annual Report, Medtronic, Inc., 3055 Old Highway Eight, Post Office Box 1453, Minneapolis, Minnesota 55440.

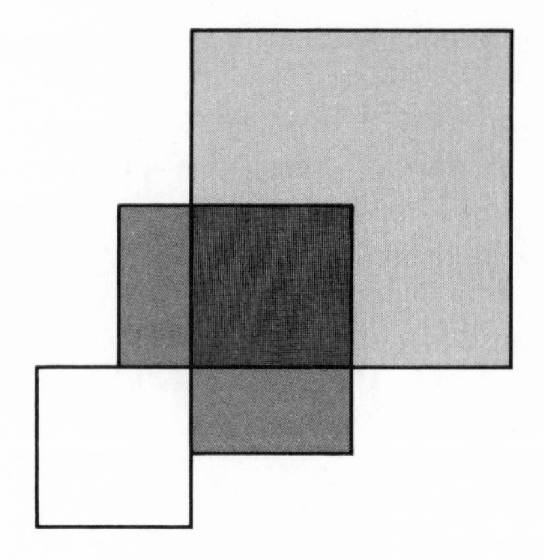

CONCLUSION

This book is a personal affirmation of technology, espousing a humanism that accepts technology as part of nature. Technology presents us with a near infinite set of possibilities which we must explore for the same reason we climb mountains — because they are there.

With the cost of computers plunging, a new age of invention is incipient. New horizons will appear in every direction, much as if ten continents were discovered simultaneously. The result will not only be new artifacts and activities but also new concepts and culture. It is possible to glimpse fragments of this future now and to sense how rich and unexpected it will be. The Responsive Environments seek to convey the essence of this vision, to make it concrete, much as a painter would try to capture a landscape.

Unfortunately, for the past century some humanists have been at odds with technologists, viewing technology as a pernicious force beyond their control — all the more intolerable because of its human origins. This attitude is part of the humanist's traditional focus on the past and unwillingness to embrace either the art or technology of the present. The effect of a work of art is strongest at the time of its creation and weakens with the passage of time. An artistic work that can be understood only with a scholar's footnotes cannot be considered more powerful than one that speaks to its audience directly. The common assumption among lay people that fifty years hindsight is required for the identification of a work of art is based on a lack of confidence, denies the value of art created in the present, and makes aesthetic judgment a kind of historic Nielsen rating. This attitude goes back to the worship of the Greeks by the Romans. It is anti-art. The young are taught to appreciate past art much as physicians expose people to a weakened form of a virus so they will become immune to the real thing.

Art exists in the present. It affirms it. We are living in epochal times, inventing technology with existential implications. Surely the art of such times must use the most powerful means of expression available, leading rather than following in the exploration of aesthetic technology. Artists cannot remain aloof, cultivating an immaculate ignorance of all things technical.

But technological art is not an established medium with a set of self-explanatory tools that one can use without understanding how they work. The tools are recalcitrant. They must be mastered if they are to submit to the will of the artist. Otherwise, artists must settle for what they are willing to do easily. Unfortunately, the tools that comprise the medium are fragmented. None of the constituent technologies approach the flexibility of a traditional medium like oil painting. Each can support only a few works before it is exhausted; therefore the artist has to be able to invent new media as well as create compositions using them.

METAPLAY, PSYCHIC SPACE, and VIDEOPLACE were aesthetic experiments. When viewed as conceptual art which sophisticates those who experienced or heard about it, these pieces were successful. The response of the public to these exhibits was good. Most of those who experienced them were excited. Some were moved. A dimension people had never considered had been opened to them.

The idea of the Responsive Environment identifies a new aesthetic option, an important new way of thinking about art. Currently, it lacks the visual impact of film, the narrative capability of the novel, and the complex expectations of music. Ultimately, it should be possible to match these art forms in those dimensions. But the success of the Responsive Environment lies in the importance of physical interaction. It depends on the discovery of new sensations and new insights about how our bodies interact with reality. It depends on the quality of the interactions that can be created.

Perhaps, in the long run, responsive technology will not communicate as art, but merely entertain, either as the quintessential video game, or as a participatory adventure akin to film. However, the intensity, the absolute focus, of a child who has entered the world of an electronic game, demonstrates the power of the medium. It must be possible to expand that world beyond beeps and explosions to include the unfolding of concepts and the communication of philosophy. Perhaps not. But it is a worthy experiment.

Another purpose of the Responsive Environments is to express technology itself, to make an aesthetic statement of the majority faith in technology, science, and human invention. We need that faith, for our achievements have propelled us through a discontinuity. Our innocence is lost even if we were to stop progress. We have begun to violate a host of old taboos: creating artificial organs; exploring space; manipulating genes, including our own; and artificially duplicating the mind. We are stripping ourselves of whole layers of mythology. Not only has nature as a godlike force been all but banished from this planet, but we now find ourselves acquiring powers once reserved for the gods. We find ourselves not the final goal of evolution, but its conscious agents.

Is this hubris? I think not. It is far more humbling to realize that we are at a beginning we do not understand very well, for which we are not prepared, and for which there is no place to go for guidance, than it is to believe that we represent the attainment of the final goal of evolution. We are beginning to understand some of the simplest laws of nature. We can see how these laws are manifest in the creatures and environment around us. But we are also discovering that there is a complementary process that may be even more challenging. Once we have apprehended these laws, we must see how they can be used in new ways. We move from analysis to synthesis. Rather than dissect, we create. In this process we discover a new set of laws that govern how things can be put together. These laws, the ones that govern how a computer or car can be built, are as natural as gravity. We like to say we invent them, but they are there, waiting to be discovered.

As our competence in shaping our environment grows, we discover that our relationship to it has changed. We are shocked to discover that our actions dictate most of what happens on planet earth. For better or worse we are faced with the fact that we must foresee the ramifications of every action or be responsible for the consequences. This is not a naturalist argument, but an engineering one. We are now charged with the design of our entire physical environment. Certainly, we will choose to preserve what we can, but the status of what is saved has already been changed forever. A preserve can never be the same as a natural habitat. A protected species is no longer a natural one.

There is no denying that this future will change what we are and how we think of ourselves. However, we cannot let fear of our own capabilities cause us to bury our talent. It is as idolatrous to fear the Golden Calf as it is to worship it. We must allow ourselves constantly to be redefined by our own actions and creations. There will be mistakes — big ones. They lie down every path we might choose. Technology has not made us happy. Nothing ever has or ever will do that. Our nervous systems have evolved to keep us restless. But before we turn our back on our achievements, we might pause to reflect. Before we change direction as so many propose, let us ask about the unforeseen consequences of that action. We have been living in a Golden Age. The economic environment we have created is every bit as delicate a web of interdependencies as the natural ecology. Once dismantled, it is unclear that it can be put back together.

Beware of those who are afraid of making mistakes and who suggest that poverty is the only virtue. If we accept this argument, we will surely achieve a state of grace, but we are going to hate it. Such modern-day nihilists are dissatisfied with anything that is imperfect; therefore, nothing that exists can satisfy them.

The products of technology are all this country has to sell. If a product can be made elsewhere, it can be made better and cheaper. The United States can survive economically only by exploiting and strengthening our leadership in all forms of technology. During the past ten years our technological superiority has dwindled. In crucial areas our edge, which used to be measured in years, has vanished. Unless we renew our commitment to progress, our day is over. We will have to be content to sit on the sidelines, admiring the achievements of others.

It is as if Apollo 11 truly frightened us. Since that achievement, we have turned our back on ambition and vision, turned against our own strength. We were persuaded that money for space exploration would be better spent on our problems on earth. For ten years we have done that. It has not worked. But by focusing on our problems, we perceived more and more problems. We noticed that every action had negative as well as positive results; we have not yet learned to balance the greater against the lesser. Instead, we are paralyzed. Those who act, those who do, those who produce in any area are maligned, while those who find fault or express fear are revered. The most powerful phrase has become: "There *may* be negative consequences."

As a person moves across a room they make many small errors that are reported by the visual and kinesthetic senses. Corrections are smoothly incorporated into the next action. As a society, our feedback mechanisms are distorted. What should be routine midcourse corrections are reported with hysteria. In the act of walking, if a momentary loss of balance were agonizing or a step too far to the right gave us convulsions, we would fall on our face and ponder whether we should ever walk again. That is what has happened. We have exaggerated our real pain and react to hypothetical pain as if we feel it.

It is not that our problems are insurmountable. On the contrary, it is that we have become increasingly naive about reality itself. Any engineer knows that it is not possible to have it all, to optimize simultaneously the speed, cost, size, reliability, maintainability, and power requirements of a circuit. It is necessary to decide what desirable characteristics to forego in order to have others that seem more important to the final application. The alternative is to do nothing. (But that is not really an alternative for the hypothetical engineer, who would be fired.)

The situation with the American people is exactly parallel. We have decided all actions lead to pain and so none should be taken; the changes should be controlled or control themselves. Soon we will discover that failure to act, decide, or change is the greatest decision of all. Inaction also hurts. Zero is a number, black is a color, and silence is a sound. At the present, none of our values will give. None of our principles can be compromised. We have two alternatives. One is to flip out, to go crazy. We are doing that. The other alternative is to decide, like the vegetarians in Samuel Butler's *Erewhon,* who were persuaded that plants also have souls, to drop all pretensions to absolute virtue and resume eating meat. We can get our hands dirty, make a few mistakes, and make compromises, secure that for all we know what we are creating is the best of all *possible* worlds.

Finally, the future need not be approached so seriously. In our actual experience of the future, epic issues will be in the background and day-to-day events paramount. To fully explore and enjoy what we are about to create will take more than practical problem solving. To truly master our tools we will have to use them for aesthetic expression, whimsy, and play. We must do this if we are to discover what it is, that what we have made, makes us.

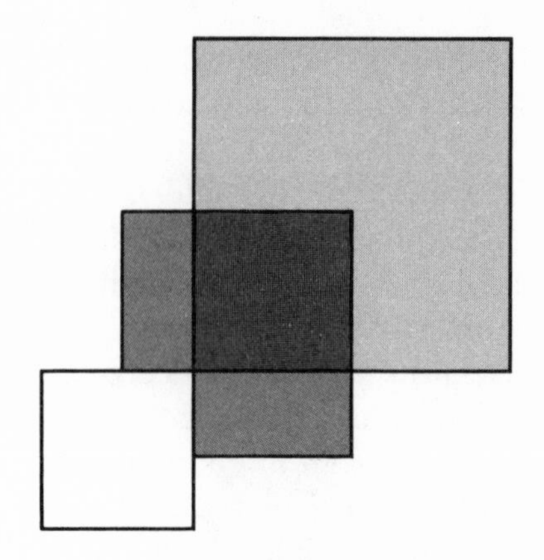

APPENDIX I DIGITAL TECHNOLOGY

When we discuss computers we are invariably referring to digital computers. We are starting to hear about digital audio recordings and even digital video. It seems the whole world is going digital. What is digital and what is so good about it?

Digital is to be distinguished from analog. An analog signal is continuous which means that if it goes from level A to level B, it must assume every value in between. In contrast, a digital signal has only two meaningful values, zero and one (Fig. I.1). In theory, none of the values in between is defined.

The physical world is analog. All of the forces we experience, such as heat, light, speed, and voltage vary continuously. Even what we call digital signals are only continuous signals trying to look digital. It does take some time to go from zero to one.

The advantage of the digital approach becomes apparent when we consider that there is often some extraneous signal mixed with the signal we are interested in (Fig. I.2). With

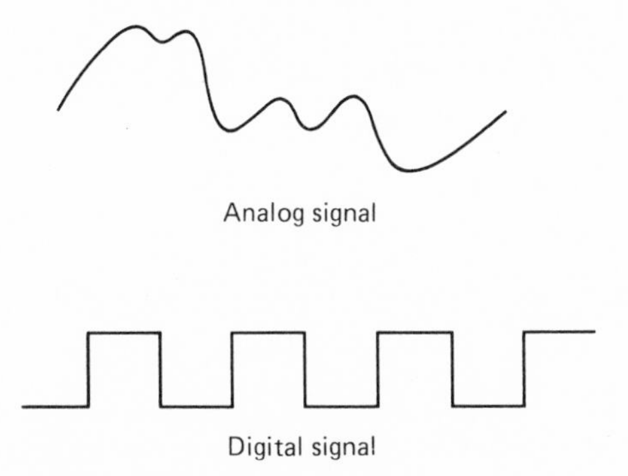

Fig. I.1 Analog and digital signals

an analog signal, this noise looks very much like the signal itself. While it is possible to suppress some types of noise by adding extra circuitry, the problem is ultimately unsolvable. Now, consider a noisy digital signal (Fig. I.3).

It is easy to design a circuit that will ignore this noise. Two thresholds are used. Below the lower, the output signal is a zero. Above the higher, it is a one. The signal spends as little time as possible in between the two thresholds. This circuit is part of every digital device. Thus, at every step, the original digital information is reconstructed whereas each successive step in analog processing further degrades the original signal.

It is this ability to reconstruct the original information that has allowed digital techniques to revolutionize electronics. It has clear advantages for any form of information storage, transmission, or processing.

Fig. I.2 Noisy analog signal

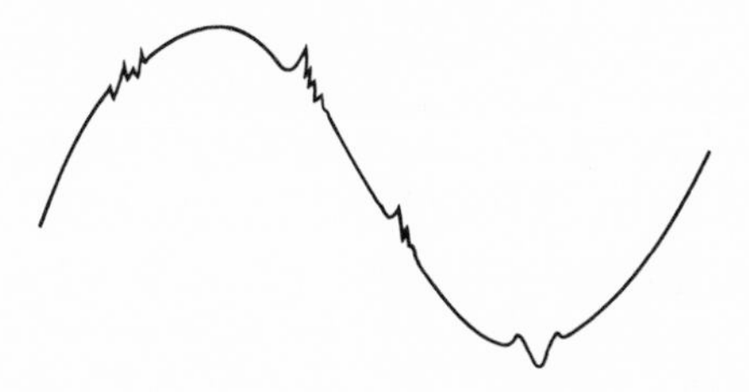

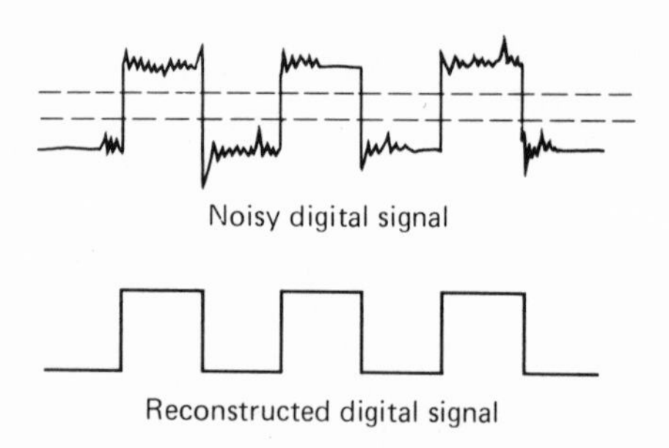

Fig. I.3 Noisy digital and reconstructed digital signals

However, a digital signal obviously does not contain as much information per unit time as an analog one. Therefore, several wires are required to carry the equivalent of even a simple analog signal. A voice signal may require eight wires; a high fidelity audio signal, sixteen. At any given moment, a wire contains one "bit" of information, i.e., a zero or a one. To create a set of digital signals from a single analog signal, an analog-to-digital converter is used to sample the analog signal at periodic intervals. The result is a step function that approximates the original waveform. Below only two bits are used to approximate the analog signal (Fig. I.4). By using more bits, a better approximation can be achieved.

Fig. I.4 Analog waveform and two-bit digital approximation of analog waveform

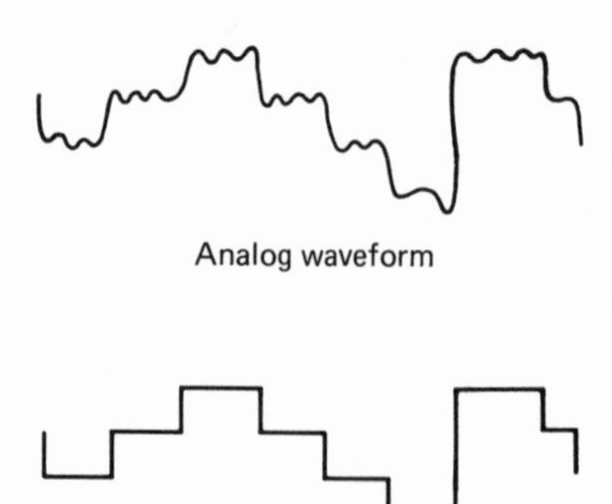

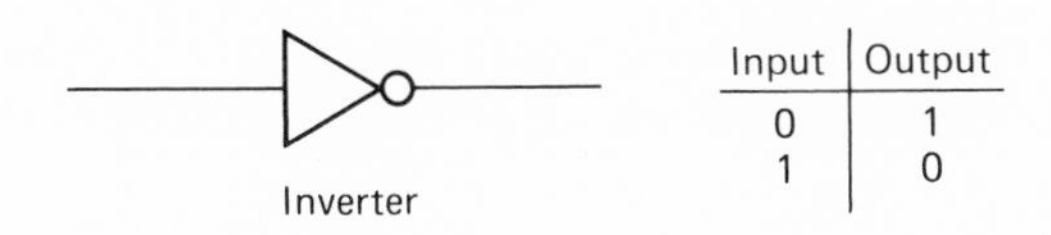

Input	Output
0	1
1	0

Figure I.5 Inverter

Logic Processing

Given signals that can be only zero or one, it may seem a giant step to the design of computers that can be programmed to be intelligent. Surprisingly, this is not so, for while there are very few functions that can be applied to binary information, very few are required. Consider one bit of information. There is only one possible operation that can be applied to it, i.e., to reverse or "invert" its value, to make a zero into a one or a one into a zero. The device that accomplishes that information is called an inverter. Its symbol and truth table are shown above (Fig. I.5). When there are two wires, several additional functions are possible. These are very simple and have intuitive names (Fig. I.6). Notice that output C is one only when input A *and* input B are true (Fig. I.7).

If A *or* B is one, then output C is one. There are also functions for "not and," "not or," and "exclusive or," which is the case when A or B is true, but not both (Fig. I.8). Of course, it must also be possible to store one bit of information. The device for doing this is called a flip flop (Fig. I.9). The value of the input is stored in the flip flop and appears at the output whenever the clock goes from zero to one. Then the flip flop will hold that value until the next zero to one transition, even if the value of the signal at the input changes. Where there are several bits of data on separate

Figure I.6 And gate

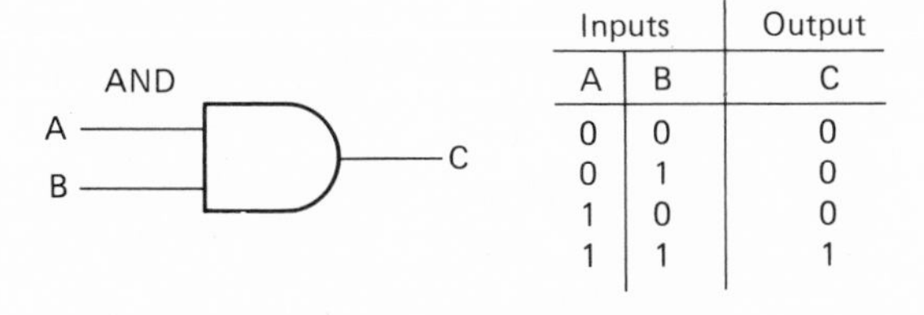

Inputs		Output
A	B	C
0	0	0
0	1	0
1	0	0
1	1	1

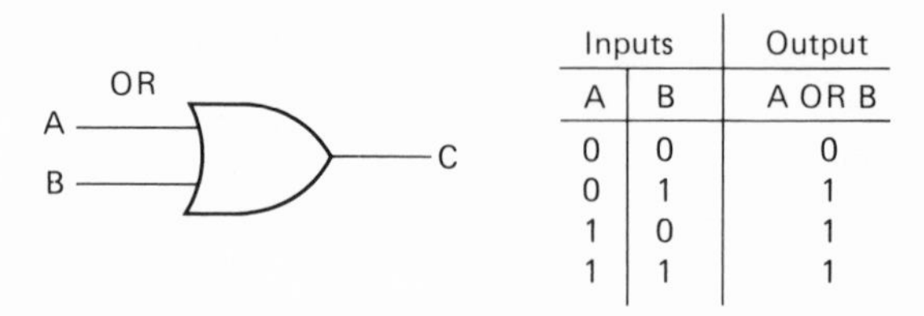

Inputs		Output
A	B	A OR B
0	0	0
0	1	1
1	0	1
1	1	1

Figure I.7 Or gate

wires, several flip flops can be used together to form a register (Fig. I.10).

Any particular computer is distinguished by the number of bits that it operates on. For example, one can speak of eight-, sixteen-, or thirty-two-bit computers. The instructions inside the computer are represented by binary patterns just as the data is. Part of the instruction will tell the computer what operation should be performed. The rest of the instruction tells the computer what data the operation should be applied to or more likely how to find it.

Figure I.8 Nand Nor gates

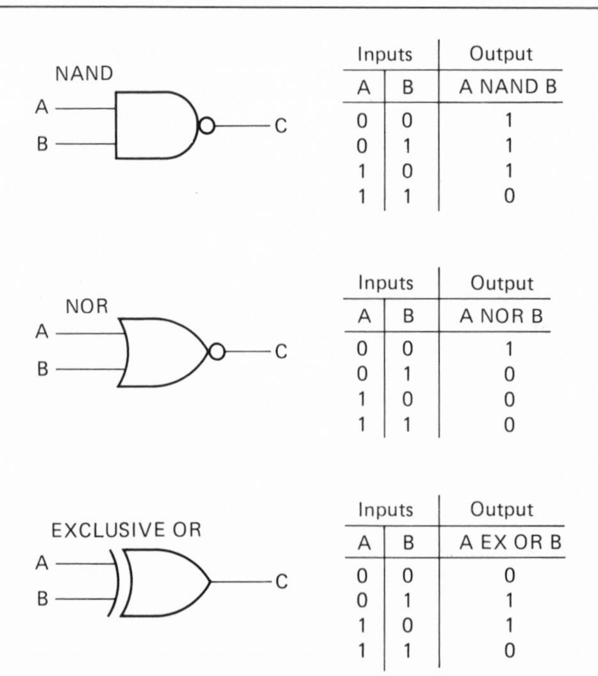

Inputs		Output
A	B	A NAND B
0	0	1
0	1	1
1	0	1
1	1	0

Inputs		Output
A	B	A NOR B
0	0	1
0	1	0
1	0	0
1	1	0

Inputs		Output
A	B	A EX OR B
0	0	0
0	1	1
1	0	1
1	1	0

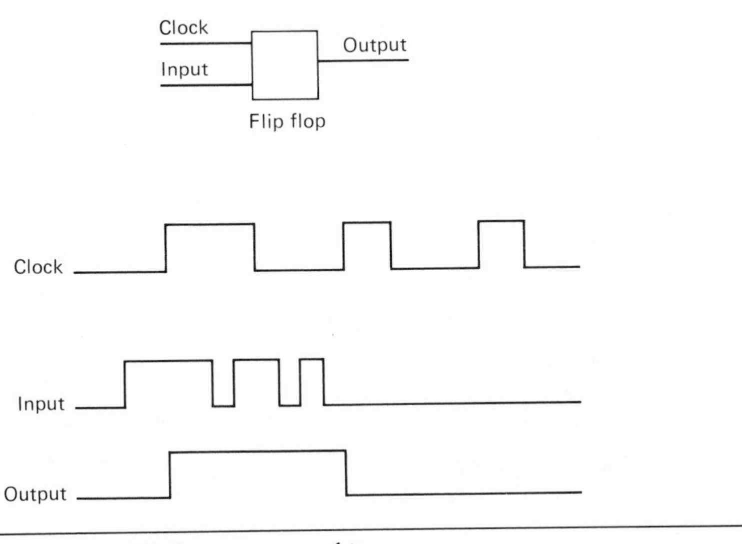

Figure I.9 Flipflop stores one bit

The primitive devices mentioned above are more than sufficient to implement any processing function available on any computer. So, in spite of the computer's reputation for being mathematical, it is really only a logical device. Because computers themselves are implemented in terms of simple logic functions, there are very simple primitive acts that you can ask a computer to perform. You can ask it to perform some simple logical function on the corresponding bits of two computer words, e.g., AND, OR.

Figure I.10 Register stores eight bits

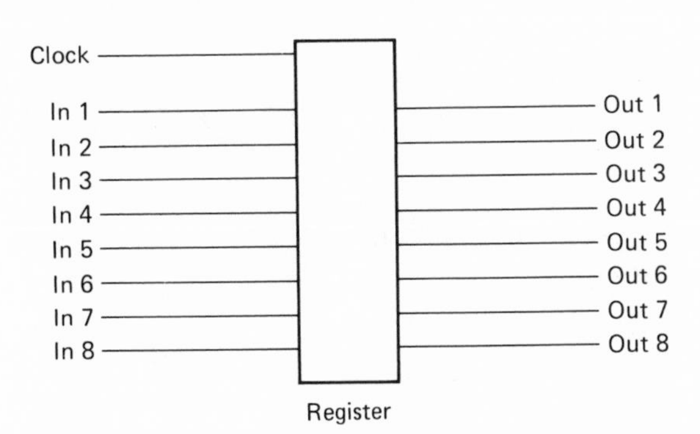

A	10110111
B	00101100
A.AND.B	00100100

A	10010111
B	11000001
A.OR.B	11010111

In fact, the only mathematical operation that many computers can do with their hardware is to add two numbers. To do this, they use binary arithmetic which is implemented in terms of simple logic functions. We are accustomed to a decimal number system. While we seldom think about it, we are all aware that the mumber 8943 really means 8000 + 900 + 40 + 3 or $8 \times 10^3 + 9 \times 10^2 + 4 \times 10^1 + 3 \times 10^0$. Similarly, the binary number 1101011 means $1 \times 2^6 + 1 \times 2^5 + 0 \times 2^4 + 1 \times 2^3 + 0 \times 2^2 + 1 \times 2^1 + 1 \times 2^0$. The binary version of $1 + 1 = 2$ is $1 + 1 = 10$ which can be computed by a simple circuit that generates the correct sum for A + B whatever the values of A and B (Fig. I.11). A somewhat more elaborate circuit must be repeated for each bit when larger binary numbers are added. The additional complexity is required to process the carryover from the preceding digit.

Figure I.11 Simple circuit adds two binary signals

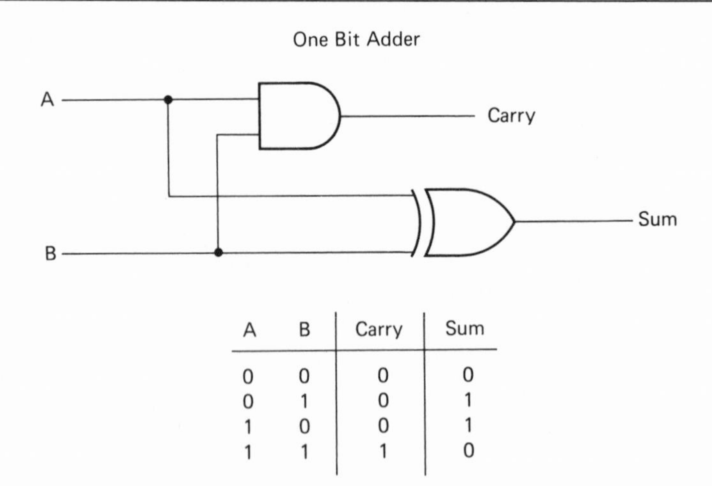

A	B	Carry	Sum
0	0	0	0
0	1	0	1
1	0	0	1
1	1	1	0

Besides these logical operations and addition, there are only a few other basic operations in computers. The computer can move data from one memory location to another. It can compare the contents of two locations to see if they are equal or if one is greater than the other. It can decide to execute some instruction other than the next one based on such a test. That is it. There is no more that a computer can do.

What makes a computer useful is the fact that it can perform these simple steps very rapidly so the final result can be very complex. Now it would be very tedious to specify a sequence of a million steps, so techniques exist for using the same instructions over and over. The first of these is called a loop. The computer is instructed to perform a certain sequence of instructions a number of times or until some condition is reached. An alternative scheme is a subroutine which is a sequence of instructions that can be invoked at any point, executed and have control returned to what was the next sequential instruction. A subroutine, then, is a form of intellectual shorthand that clarifies human thinking as well as saving time in the computer.

In summary, the greatest advantage of digital technology is that it can be used to store, transmit, and process data information without degrading it. This reliability means that for the first time in history, we can contemplate with confidence calculations involving millions and billions of processing steps. The basic digital functions are very simple. They would be of little interest if they could not be performed rapidly. The fact that logic devices can operate a billion times a second means that computers can perform their operations ten million times a second. At these speeds, even simple operations can be used to create enormously complex algorithms that produce interesting results in an acceptable amount of time. For a thorough discussion of the relation of digital technology to computing, the reader is referred to *Digital Networks and Computer Systems*, by Taylor Booth.[1]

NOTE

1. Taylor L. Booth, *Digital Networks and Computer Systems* (Wiley, 1978).

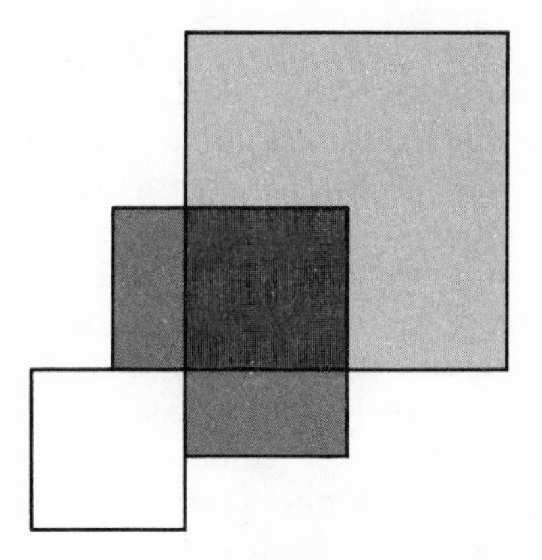

APPENDIX II VIDEO

A close look at a television screen reveals that it is composed of scan lines. The image is produced by an electron beam that strikes a coating of phosphors on the inside of a cathode ray tube. The phosphors are excited by the electron beam and emit visible light. The more intense the beam at a given point, the brighter the spot appears. The electron beam is steered by a pair of electromagnets which are coiled about the neck of the tube. This description can be confirmed by holding a powerful magnet to the surface of a television screen. The magnet will noticeably bend the scan lines, distorting the image.

The coils that control the position of the beam are fed deflection signals which cause the beam to scan the screen in a pattern called a raster. The waveform controlling both horizontal and vertical deflection is called a sawtooth (Fig. II.1). The horizontal sawtooth varies much more rapidly than the vertical. In fact, the vertical is at 60 hz whereas the horizontal is at 15,575 hz. At any point in time, the vertical

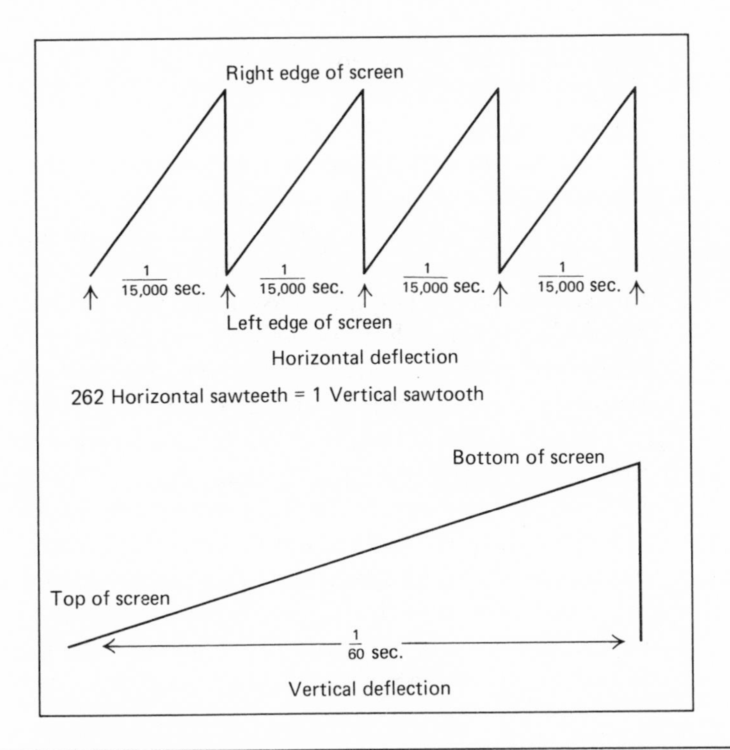

Fig. II.1 Video deflection signals

is effectively fixed at some elevation on the screen. The faster horizontal sawtooth causes a horizontal line to be drawn from the left edge of the screen to the right edge at the elevation defined by the vertical deflection. Thus, a single scan line is drawn.

During the drawing, the intensity of the beam is varied to produce the details of the image. While the line is being drawn the vertical deflection is increasing slowly so the line is actually at a slight slant. After the peak of the sawtooth has been reached and the beam returns to the left edge of the screen, it is a little lower than before and ready for the next scan line to be drawn. To prevent this retrace from being seen, the beam is blanked, i.e., its intensity is reduced to zero, until it is back to the left edge of the screen. This horizontal scanning process is repeated over and over until the beam reaches the bottom of the screen. Then the beam is again blanked as it is moved to the upper left corner of the screen.

The raster pattern controls the creation of the video image by the camera and its display on a television screen. The signal that is generated varies in only one dimension, amplitude, and yet is used to control the display of a two-dimensional image. For this reason, the video signal includes synchronization pulses as well as brightness information. These are used to assure that both camera and monitor are scanning identical points on the screen at all times.

While it would seem natural for a television to scan the first line, then the second, and so on, that is not what is done. Instead, the odd-numbered lines are scanned consecutively and only then are the even-numbered lines displayed. This practice is called interlace and is employed on all broadcast television (Fig. II.2). If this practice were not employed, we would be aware of the scanning process itself. Because the phosphors at the top of the screen would decay before those

Fig. II.2 Interlaced scanning

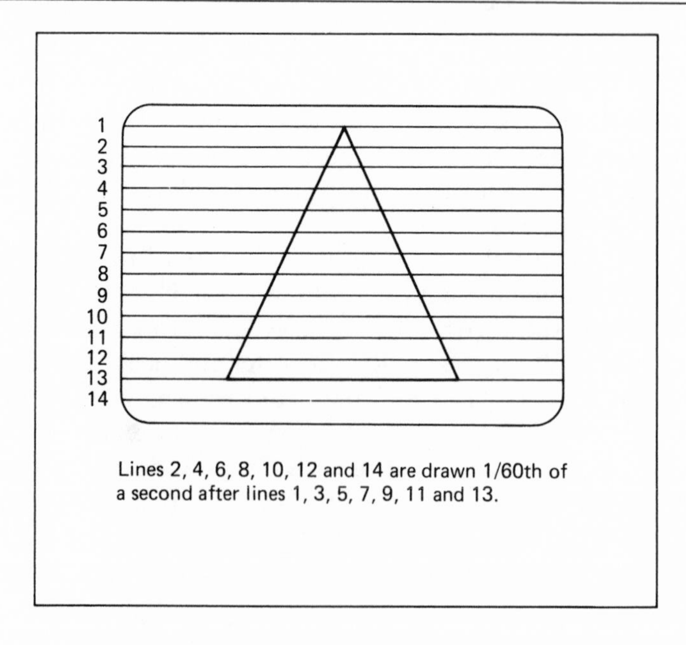

at the bottom were drawn, the screen would appear to flicker disturbingly. With interlace scanning, the first line does fade before line 480 is drawn but we do not notice because line 2 is still glowing brightly, having been drawn only 16 milliseconds earlier as opposed to 33 milliseconds for line 1.

Color television is based on a nightmarish collection of engineering and human-factors tradeoffs dictated by the technology of thirty years ago. A color image is generated by a camera with three tubes, one each for red, green, and blue. (Artists note that when adding light the primary colors are red, green, and blue. When mixing pigments, which absorb light, the primaries are red, yellow, and blue.) In a color television set three electron beams scan together, exciting red, green, and blue dots on the face of the tube. In between camera and monitor the three color signals are combined into a single composite video signal, which must be compatible with black and white as well as color sets. This compatibility is achieved by creating a luminence (brightness) signal for black-and-white sets that is a weighted sum of the three primaries. The color information itself is encoded in a higher-frequency signal, which is superimposed on the black-and-white information. This chrominance signal is both phase- and amplitude-modulated and a function of only the red and blue information. The green component is reconstructed comparing this information with the luminence signal.

In the scan-modulation techniques mentioned in Chapter 7, the image is distorted by substituting synthesized waveforms for the sawtooths that are standardly used to create the raster. In vector graphics the deflection principles are the same except that there is no scanning structure like the raster. Instead, the beam is directed wherever lines are to be drawn. If the beam is to be moved without drawing a line the beam is blanked until the desired position is reached.

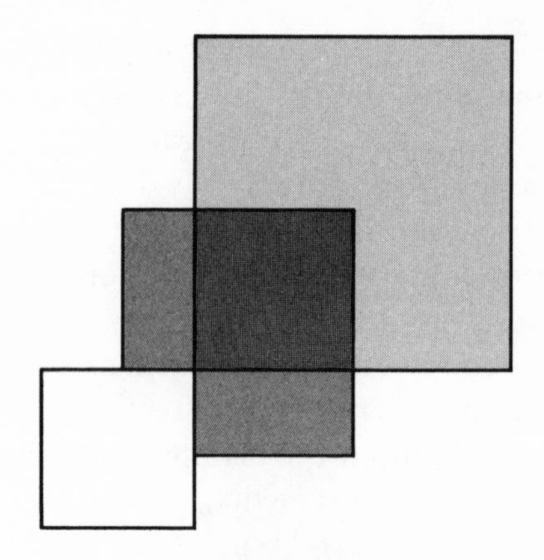

APPENDIX III ELECTRONIC SOUND

Between a stereo amplifier and a speaker, sound exists only as an electrical signal. Electronic music is based on the observation that any electronic signal, in the right frequency range, can be used to produce a sound if it is amplified and fed to a speaker. Therefore, a variety of electrical signal generators have been tried as the basis for analog sound synthesizers. Unfortunately, the basic waveforms that are most easily produced with analog circuitry are quite simple. Typically, only four waveforms have been provided:

Amplitude and frequency of these basic waveforms can be controlled either by knobs or by other waveforms. An amplitude envelope for individual notes can be defined in terms of attack, fallback, sustain, and decay. The attack is the time it takes a note to reach peak volume. Fallback is the time required to settle to full volume. Sustain is the time it holds full volume. Decay is the time it takes the sound to fade away at the end.

Since the basic waveforms are very simple and very boring by themselves, two methods are available to create more complex waveforms from the basic four. The first is to use the output of one waveform generator to control the amplitude or frequency of another. The second is to take advantage of the fact that three of the basic waveforms are not basic at all. They are the sum of a number of sine waves of frequencies of equal or higher frequency. A series of filters are usually provided that will eliminate all frequencies that do not meet a certain criterion. For instance, there are high pass and low pass filters that pass all frequencies above or below a certain frequency. Band pass filters allow only frequencies within a certain range of a center frequency and a notch filter passes only what a band pass filter rejects. Note that the control frequencies for these filters can be set by another waveform.

In the past few years, digital synthesizers have begun to appear. Some of these are very simple and are designed primarily for games. Others are very sophisticated and are capable of producing any sound. Digital sound synthesis generates sounds by producing a series of digital samples whose values approximate the desired waveforms when converted to analog voltages.

Since the waveform is an approximation based on a finite number of samples, the sampling process itself may produce unwanted audible effects. For this reason, the sampling frequency should be at least twice the highest frequency which is to be generated. This requirement, which is called the Nyquist criterion after the man who originally stated it, means that digital audio systems must operate at a minimum of 40,000 samples per second.

The computer can generate sounds itself by driving a digital-to-analog converter (DAC) directly. Of course, just putting out the 40,000 values per second required to produce undistorted sound leaves the computer very little time for deciding what sound to produce next. It seems desirable to delegate the actual generation of sounds to some kind of external waveform generator. The synthesizer used in the VIDEOPLACE system is representative of digital sound techniques. It is based on Fourier synthesis of complex waveforms.

The Fourier theorem states that any periodic waveform can be represented as the sum of harmonically related sine waves. The base frequency that we think of as the pitch is called the fundamental. The frequency of the first harmonic is twice the frequency of the fundamental. The second harmonic is three times the fundamental, and so on (Fig. III.1).

The VIDEOPLACE synthesizer is designed to generate and sum sixteen sine waves which may or may not be harmonically related (Fig. III.2). The programmer has control of the frequency, amplitude, and phase of each of the separate

Fig. III.1 Summing harmonics to approximate a square wave

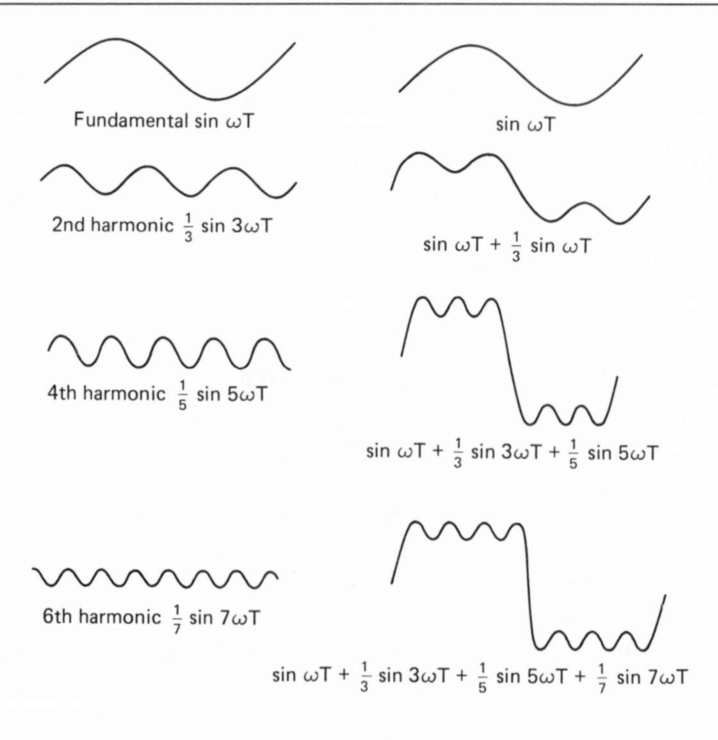

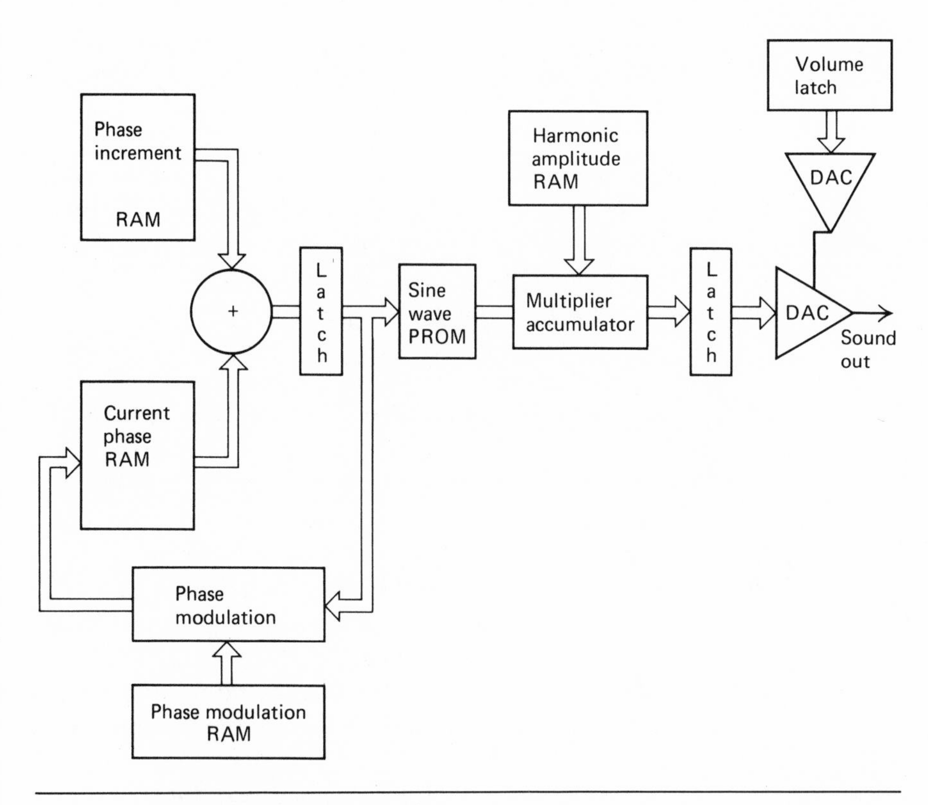

Fig. III.2 Digital sound synthesizer

sine waves as well as amplitude control of the sound as a whole. The device generates a sound sample for each of the sine waves and sums them 40,000 times a second. Since all of the waveforms are sine waves, a single lookup table with digitized sine wave values is sufficient. At any moment, there is a current phase angle for each of the waveforms. Their pitch is simply an increment that is added to the current phase angle each sample period. The larger the increment, the faster the sine wave table is traversed, the faster the waveform repeats itself and the higher the frequency.

The sample cycle is divided into several stages. During the first, the current phase angle of one harmonic is added to its phase increment to produce a new current phase angle which is stored. In the second, the sine of this new current phase angle is looked up in the table. In the third, the sine wave sample is scaled by the current amplitude for the harmonic and added to the sum of the harmonic samples so far. When all sixteen harmonic samples have been summed, the sample is fed to a multiplying digital-to-analog converter that creates an output voltage scaled by the overall amplitude. These four steps are run as a pipeline so each of the hardware stages is in continual operation.

The advantages of controlling sounds as the sum of waveforms are, first, any sound can be produced, and second, a single framework for controlling the sound is created. The use of voltage-controlled filters in analog synthesizers is unrelated conceptually to the rest of the sound generation process. With harmonic generation, the effect of filtering is gained by explicitly adding or deleting waveforms. This allows effects that are impossible with physically realizable filters. It also allows far more precise definition of timbre, because the perceived differences between instruments are far more a function of the relative attack-and-decay patterns of the various harmonics than of the harmonic content of sustained notes.

The original version of this device was built in early 1976. There is an irony associated with its use. While it is powerful on its own, it requires considerable computer power to control it and considerable disk storage to keep the control information that is used to define higher level sounds. Thus, its full exploitation in the system requires the dedication of an additional processor to its control. Similar devices have been built at Bell Labs and the University of Pennsylvania.[1]

NOTE

1. H. G. Alles, "Music Synthesis Using Real Time Digital Techniques," Technical Report, Bell Laboratories, Murray Hill, New Jersey.

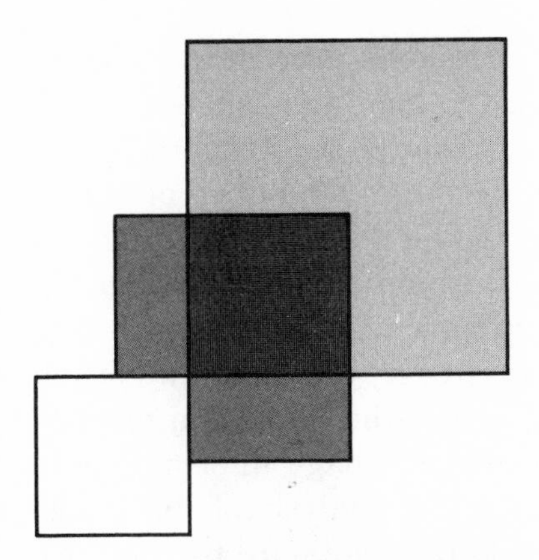

APPENDIX IV IMAGE SENSING

Pattern recognition is a branch of computer science with an extensive literature and a large inventory of techniques. Most of this effort is devoted to complete and correct analysis of complex scenes with no time constraint. In the Responsive Environment, it is accepted that perception will be imperfect. There will be moments when the computer will be confused, just as there are moments that confuse the human when events are changing rapidly. Rather than wait until the pattern recognition issues are totally resolved before implementing a real-time system, the approach taken in this research is to sketch out a perceptual system that performs in the required time frame and then to chip away at the problems so the competence of the system is gradually expanded. While on the one hand, this is an ad hoc approach, it reflects a theory of vision and intelligence in general that argues that we have an enormous number of very specific operators that we can employ very rapidly once the correct domain has been identified. Also, while we undoubtedly have more formal models of articulated three-dimensional forms than we can use to reason about what we see, this process is relatively slow and not part of high-speed human performance.

The processing required to analyze an image in a Responsive Environment is divided into a number of stages that can be thought of as happening sequentially, although events at later stages will influence future processing at the earlier ones. The stages are: the processing of the grey-level image to create a high-contrast image, the digitizing of the participant's outline, the analysis of the outline to identify salient features, the use of previous image information to help identify confusing features, the analysis of the image with respect to graphic objects to recognize effective actions, and the evaluation of sequences of movements to identify gestures.

The first stage can be very simple or quite elaborate. The simplest approach is a simple threshold applied to the grey-scale image. Any grey level above the threshold is white and any grey level below is white.

Due to the nature of the video signal, there are a number of problems with this approach. First, the video signal always contains noise. The noise means that a given point will be above the threshold in one frame and below in the next, on and on. Thus, edges will appear to move back and forth continually. Another problem is that video monitors contain an automatic gain control circuit that tries to maintain an average level of brightness for the image as a whole. Thus, if the whole image should go to black, the displayed result will be grey.

The problem this causes is that a change in brightness in one part of the image can cause the grey level at another point to change.

Yet another problem with a fixed threshold is that the sensitivity of a television camera is not uniform throughout the visual field. Given a uniform input signal, a camera will generate an output that shows small gradations of brightness around the screen. One method of compensating for these problems is to detect not an absolute threshold, but abrupt relative changes in brightness. Techniques exist for accomplishing this sort of processing reliably, but they are not easily accomplished in real time. While it is possible to get a useful signal, better preprocessing would serve to improve the resolution.

The next step is the processing of the high-contrast image to find the participant's outline. There are a number of ways this can be done. The simplest is the one currently taken which is to just save the first and last nonbackground point on each line. Another approach is to filter the image using nearest-neighbor techniques. This processing can ignore isolated points or those features that are not likely to be part of a well-formed outline. Another function that can be applied is local smoothing which can generate an outline with greater apparent resolution than the original sampling. This smoothed outline will be less sensitive to noise because the effect of isolated spurious points will be averaged with the good neighboring points.

Once the outline has been determined, it is scanned for important features. The left and right edges are searched for peaks that will indicate the presence of hands, elbows, feet, and knees. Additional features are horizontal discontinuities that indicate the shoulder or an outstretched hand and vertical or diagonal sections which indicate the length of arms, legs and trunk. Comparison of the left and right extremes on each line can be used to identify the head, neck, outstretched arms, waist, and legs. As long as the participant is facing the camera with his or her arms to the sides of the body, the outline extremes suffice for the identification of body parts. The only clear exception is the case where both hands are raised above the head, obliterating any information about the exact position of the head. The full outline can then be analyzed in the region of the head. Thus, more elaborate processing of the whole image is avoided. It is only applied where there is a high probability that a particular feature will be found.

The outline information may successfully identify the position of a person's head, hands, and feet on the screen, but not yield three-dimensional information about whether an arm is extended forward or to the rear. The system should be aware of such large movements. A camera mounted to one side of the participant will detect the presence of a limb extended forward or to the rear and indicate by its elevation whether it is an arm or a leg. A ceiling-mounted camera can also detect an extended limb and determine left from right but not foot from hand. More ambitious is the use of a second camera close to the original one so the two function as a stereo pair. If corresponding features in the two images can be identified, the distance between them and their general position on the screen can be used to calculate the depth of the feature in the scene. When these calculations are applied to complex grey-scale images, the calculations required to match features are very time-consuming. However, with the outline images, matching corresponding points is straightforward in many postures.

Assuming that the outline information has been sufficient to identify the parts of the body correctly, it is necessary to consider their position in each frame with respect to its significance to the current interaction. The processing at this step will not always be the same. The rules of the particular interaction, the immediate graphic context, and the previous movements of the participant will determine which aspects of the participant's current posture are important.

It may be important to know whether the participant is touching a particular graphic object. Since the participant's one-bit, high-contrast image can be directly compared to the image of the graphic object, the hardware can digitize the area of overlap. This information can then be used to control the movement of the object if appropriate.

It is also of interest to note whether the participant has moved and, if so, how much and in which directions. Again, we have hardware to directly compute some of this information. The current image and the one immediately preceding are logically compared and the differences digitized. By checking the width of the different areas, the speed and direction of movement can be inferred. Features that lie in the areas of difference are obviously the ones being moved.

When several frames are compared, trajectory information can be derived. Also, the semantics of a particular motion can be understood by considering previous movements. For instance, if the participant's hand is found to be in contact with a graphic object, analysis of previous frames will determine whether the contact was the result of a touch, a tap, or a hit. Similarly, if the participant has been carrying an object, the movement of the hand must be considered to determine whether the object will fall.

There will also be cases where the program will interpret pure gesture information as opposed to analyzing actions. A particular sequence of hand movements may be given a special significance. For example, the participant could use his or her finger to draw a letter that would be recognized and acted upon by the computer. In this case, what is analyzed is not a single perception, but a sequence. At any moment, the recent string of perceptions is compared to a set of meaningful gestures. At any moment, the entire sequence may match a particular gesture. More likely, the most recent sequence partially matches several of the gestures which are then considered active. It may also completely eliminate previously active hypotheses from further consideration.

At any of the points above, the information found at one level may be insufficient to make a determination. When this occurs more elaborate processing must be employed. For instance, when a person's hands are in front of the body, it is impossible to be sure of their position from the outline sensor alone. To make this determination, more sophisticated processing can be applied locally under the guidance of the information gained from the outline. Special operators may search for edges within the grey-level image of the person's body. Earlier views of the participant's image may have revealed information about the orientation of textures within the participant's clothing that can be used to identify their forearms. With a color camera, the color of the hands may be readily found against the background of clothing. If the participant has been moving, considering those movements may help to constrain the possible locations of their limbs and thus guide the search for them.

With all of these techniques, there will be times where the input is ambiguous. The computer will either have to choose what it considers the most likely alternative, or adopt a coping behavior that minimizes the damage the momentary confusion causes. While it is possible that subsequent frames will shed light on the confusion, the system's interest in the past is very short-lived because there is a new moment to be dealt with every one-thirtieth of a second for both the person and the machine.

Hardware

There are many aspects of the image that can be analyzed to provide useful information about participants and their behavior. While the answer to one question may make asking another unnecessary, in a parallel-processing environment where time and not computing must be conserved, it is possible to ask a number of questions in parallel. Some results may serve only to confirm others, or they may be overruled and ignored. To facilitate this feature, all video signals are available on the bus to allow a number of specialized processors to analyze the video image simultaneously. There is currently one such bus dedicated to high-contrast images and outline processing. The processing on this bus does not store the current image and then analyze it from a memory. Instead, the outline or outline extremes are detected and processed on the fly.

A second bus is designed and not yet implemented. This bus would be designed to facilitate nearest-neighbor processing. The circuitry would store several previous scan lines and present pixels and their nearest neighbors on the bus simultaneously at video rates. Then, a number of programmable feature detectors can be attached to this bus. When these devices detect the condition to which they are sensitive they too can put a synchronized signal on the bus. Higher-level detectors can then look for meaningful clusters of such features to identify specific parts of the body.

Assuming we had such circuitry, we could simultaneously compare each point to its eight, sixteen or twenty-five nearest neighbors. With special-purpose hardware, edges in different orientations would be characterized by different relationships among the eight neighbors and the given point.

Considering only the 3×3 case for the moment, six bits per pixel require fifty-four lines on the bus. Special circuits can be constructed that will compare the nine points according to some criterion and generate a one when that criterion is met and a zero when it is not.

For example, suppose each point is compared to a particular threshold and a one is generated if it is above the threshold and a zero if it is below. There are nine bits, one for each point. These nine bits can be used as an address into a memory that contains a one if the combination of the nine bits meets the criterion and a zero otherwise. A line detector might then generate ones for any of the groups of pixels (Fig. IV.1) if several conditions were to be tested for, as many bits could be stored for each combination of inputs. Note that by changing the contents of the memory, the circuit could rapidly be reprogrammed to look for different features.

Fig. IV.1 Nearest-neighbor edge detector

1	0	0
0	1	0
0	0	1

0	0	0
1	1	1
0	0	0

0	0	1
0	0	1
0	0	1

0	0	1
0	1	0
1	0	0

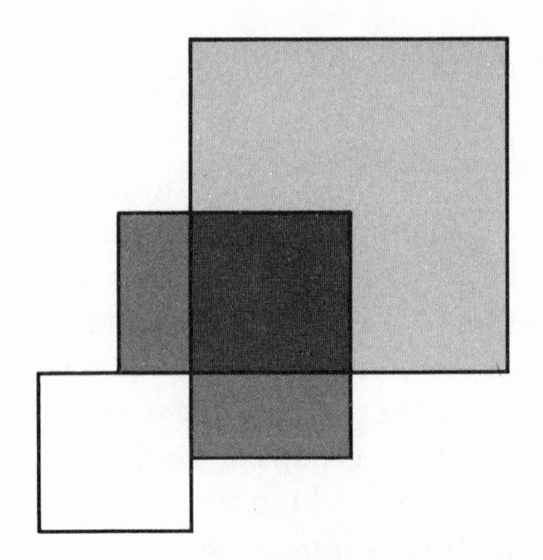

APPENDIX V
RASTER GRAPHIC ISSUES

Raster Graphics

While raster graphics hold the promise of becoming an ideal display technology, there are a number of constraints that must be described if their current status is to be understood. These constraints are based on the processing load, the resolution of the medium, and its combination with standard video equipment. The limitations are imposed by the nature of the medium. What is drawn is actually not lines, but points. If a line is drawn, or the boundary of a solid area defined, the intersection of the line with each of the scan lines it intersects must be computed. These tasks increase the computer processing time considerably. In addition, colored and shaded areas must be filled in on a point-by-point basis. Where continuous gradations of shading are required, the intensity and hue of each point must also be computed.

To get a feeling for what this responsibility for specifying every point on the screen implies, consider a 500-pixel-by-500-line display. It has a total of 250,000 pixels. Assuming roughly 1 pixel equals 1 character, 65 characters make a printed line, and 30 lines make a page, a single frame is equivalent to a hundred pages of a paperback novel. Whoever said a picture was worth a thousand words was a master of understatement. Remember too that the picture being discussed here, a video frame, is only good for a thirtieth of a second.

The pixel-by-pixel control of the graphic screen is what gives raster graphics its representational power. At the same time, it seems very inefficient in cases where an area of the screen is to be filled in with a single color, shade, or texture. In these cases, it is possible just to store the edge coordinates of the domain. During the scan, when the left edge coordinate is reached, the color is turned on and then turned off when the right edge is reached. For complex shapes, odd-numbered edges could turn the color on and even numbered edges turn it off. A similar strategy can be used when the domain outline is plotted into a one-bit-image memory. Another variation is to code the width of the domain along each scan line.

There are also a number of resolution constraints that make raster images as obviously computer generated as vector images. First, since each pixel on the screen corresponds to a memory cell, greater resolution requires more memory. In fact, doubling both the horizontal and vertical resolution increases the memory required by a factor of four. Increasing the vertical resolution has far more serious implications than increasing the horizontal, because broadcast standard television has a fixed vertical resolution of about 480 visible lines. Therefore increasing vertical resolution means leaving the well-supported world of consumer products and venturing into the more forbidding territory of unstandardized industrial products. New color systems with 1000 × 1000 resolution exists, but they are very expensive and do not yet permit projection.

It may seem that resolution is not of great concern because people watch television without any great dissatisfaction. However, television starts with an extremely complex and varied real-world analog image that is smoothed by the analog characteristics of a camera. The computer, on the other hand, generates digitized objects that are either present or absent at any point on the screen. So a line drawn on a raster display is going to be noticeably staircased on a 500 × 500 display (Fig. V.1). Even 1000 × 1000 displays have a perceptibly digitized appearance. This digitizing is particularly evident with lines that are sloped near the horizontal or the vertical. There are special algorithms that correct these effects, but these require large amounts of additional computation.

The vertical resolution is not a simple function of the number of scan lines. The video frame consisting of 480 visible lines is refreshed thirty times a second. However, each frame is actually divided into two fields with the even-numbered scan lines in one field and the odd-numbered in the other. Therefore a horizontal line only one scan line thick is refreshed only thirty times a second producing a very disturbing flicker. This effect can be eliminated by doubling the line thickness so it consists of two scan lines. Since one of these scan lines is refreshed every one-sixtieth of a second, the persistence of vision of the human eye integrates the two lines into a single one that appears steady. It is important to note, however, that the fix requires cutting the vertical resolution in half.

Fig. V.1 Staircased line on raster display

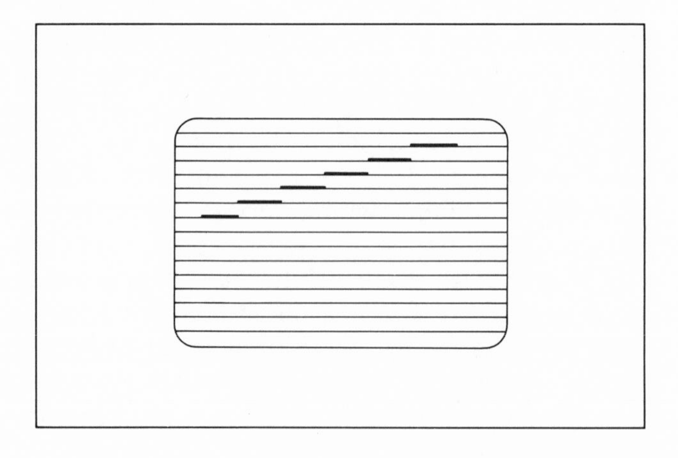

For a given horizontal and vertical resolution, there is the separate issue of color resolution. Each pixel may be represented by one or more bits of memory. The number of bits per pixel limits the number of different colors that can appear on the screen at a time. One bit per pixel means a binary black or white image. Two bits mean 2×2 or four colors. Fifteen bits mean thirty-two thousand.

Another resolution concern arises when standard color video equipment is used. Color video information is represented by the bizarre encoding scheme in NTSC (National Television Systems Committee) commercial television. This encoding scheme allows differing bandwidths for each of the primary colors and uses differing proportions of each to create the higher bandwidth intensity information. The result is that in the case of narrow vertical features the appropriate color may not appear, although the intensity information will. In the case of broader features, the color may appear displaced to the right on the screen.

Some horizontal color juxtapositions may not work because the phase encoding of the color information cannot change abruptly from one color to the next. Rather, intermediate colors from the spectrum will temporarily appear. In addition, if an attempt is made to shift too rapidly from color A to B and then to C, no color may appear at B as the command to proceed to C supersedes the command to produce color B.

Finally, in the case of repetitive texture patterns that alternate from one color to another, wild scintillation effects are created. In broadcast television these effects are compensated for by controlling lighting, makeup and dress. For example, clothing with checks, plaids, and stripes are avoided. However, in computer-generated images, sharp edges and abrupt transitions are the rule, not the exception; therefore, the world of broadcast color is not the simple one of the X,Y plotting of red, green, and blue (RGB) information that it first appears. Rather it is a complex medium where a variety of constraints must be satisfied in order to produce acceptable images.

RGB Displays

The alternative is to avoid broadcast format video and to use special RGB displays. These are currently exotic items with limited markets and higher price tags. However, the visual result is so dramatic that it is well worth the extra cost if it can be afforded. Anyone who has seen European television will have a sense of what the improvement might be like. To date, the United States has been satisfied with the world's poorest-quality video images, but the current desire of business, science, and government to display information more and more clearly will force the development of sharper, larger, and visually more powerful displays. The Japanese have recently decided on a new television standard with 1,100 lines as opposed to our current 525. The original reason for our video format was the conservation of the frequency spectrum. However, the availability of cable television and direct satellite transmission to the home will circumvent that concern.

Another problem with raster graphics is that very few manipulations are possible within the display space. With vector graphics, movement, scaling, and rotation of image elements is possible without changing their definition. However, in raster graphics, the final image is formed in the refresh buffer before it is displayed. Each element that is written into the buffer overwrites any information that is already there. Therefore, if an element is to be moved, it is not just a matter of redrawing it elsewhere. Its previous occurrence must be erased and the original background restored. If there is no background the problem is simplified. It is possible to define an image element independent of its location and to move it about the screen. However, the required hardware is different from the hardware commonly used for raster graphics; it is more akin to the architecture used for alphanumeric displays and video games, both of which move a small number of image elements about the screen.

Furthermore, in existing raster graphics systems there is no way to scale or rotate an element in the display hardware itself. While techniques exist for performing these transformations in the raster space, it is not clear whether they will be widely available in hardware products or restricted to video special effects systems.[1] The need for this development may be forestalled as the speed of computation increases to the point where complete new images can be generated in real time. In particular, the development of a twelve-chip experimental system implementing 1,344 processors specialized to the trigonometric computations involved in 3D computer graphics suggests that the problem of real-time graphics is considered important enough that its solution is inevitable.[2]

Raster graphics has the ultimate capability of producing extremely complex and convincing representational displays. However, these images require extensive computation to generate and allow only limited manipulation once they are created. The advantage of having color is offset by the quirkiness of the broadcast medium and the relative expense of the RGB alternative. Even with these misgivings, the current momentum is clearly toward this technology. (For additional information on raster graphics, see *Principles of Computer Graphics* by Newman and Sproull.[3])

NOTES

1. Carl Weiman, "Continuous Anti Aliased Rotation and Zoom of Raster Images," *Siggraph* (1980):286–301.

2. James Clark, "A VLSI Geometry Processor for Graphics," *Computer* (July 1980):59–68.

3. W. M. Newman and R. F. Sproull, *Principles of Interactive Graphics* (McGraw-Hill, 1979).

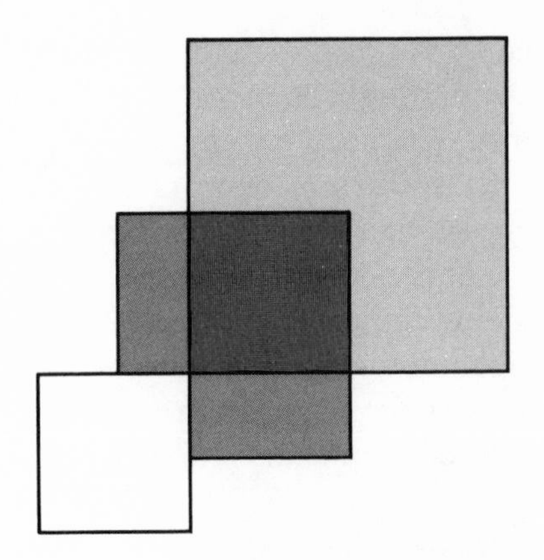

APPENDIX VI
SYSTEM ISSUES

SYSTEM ISSUES

The main thrust of computer science research in the past has been in addressing the needs of the large general-purpose multiuser system. While response time is considered, it is always the most elastic requirement. Since the human components are thought of as having no value, their time and their achievements are not even considered when judging the performance of such a system. For this reason, the computing resources available to the individual have not changed radically in the last fifteen years even though the cost of computing has plummeted. Instead, the number of simultaneous users and the total user community have increased enormously on both the large and small machines. Ironically, those of us who started with unlimited access to large machines in the mid-1960s were relegated to minicomputers in the early 1970s, and find ourselves facing microprocessors today.

Finally, the end of this process has been reached. Even in time-sharing systems, the terminals typically contain a microprocessor. This small bit of personal computer power is the vanguard of a revolutionary shift in computer systems, the dedication of at least one computer to the individual user. We can expect the expansion of this resource to be a major theme in future computer development. The Responsive Environment takes this process to its extreme, focusing an entire system composed of a number of computers on the needs of one participant.

Given the need to coordinate the operation of a number of computers toward a single goal, one must consider how they can be interfaced from a hardware and software point of view.

The literature contains descriptions of a number of experimental systems that employ parallel processing. Most of these systems are again motivated by the desire to support general-purpose computing. Even when this is the goal, there are a number of approaches that can be taken. The interconnection of multiple processors on the Ethernet system is motivated by the desire to share system resources.[1] The interconnection of the IMP processors at an ARPANET node is motivated by reliability.[2] In principle, a bullet in the brain would not hurt this system, because no one processor is in charge. Other systems such as DEMOS and MICRONET use message-oriented, byte serial communication among processors.[3,4] CM* and C.MMP are more closely coupled, with the former using a system-wide address space and the latter swapping blocks of memory among processors.[5,6] However, with the exception of the IMP all of these systems and the operating software that controls them are directed at studying general problems in parallel processing rather than the strategies required to minimize response time for a certain class of problems.

Most large systems spend a very significant percentage of their time and resources deciding who is going to be allowed to do what next and how to charge them for it. Similarly, in experimental parallel-processing systems, the fact that every processor has the potential of communicating with every other means that effort is expended deciding who should communicate to whom. With a true real-time system, which means one-thirtieth to one-fifth of a second maximum, this kind of generality cannot be afforded because the system must be prepared to act without hesitation. This means that processing, power, and communication paths must be generous. They are not critical resources whose use is optimized. On the contrary, they are potential bottlenecks that must not be allowed to impede response. Wherever possible, they must be dedicated to the function they serve, for in such systems time is the critical resource.

In as much of the system as is practical, prioritized interrupts are avoided. They generate complex timing behavior that represents an uninteresting source of confusion when testing the system. Refining performance by tuning priorities is like trying to tune a radio with forty knobs. The only way for humans to understand something quickly, easily, and without error is if it is sequential, hierarchical, and well structured. Therefore, it is desirable to make the system as much like a clocked synchronous circuit as possible. The system's hardware and software must have an understandable, easily analyzed rhythm. One of the major rhythms would be the one-thirtieth-of-a-second video frame time — at each stage in the reflex processing, all computations have to be accomplished in this amount of time. The actual response to a given action might go through several steps of processing, each requiring one-thirtieth of a second. The result of each step would be passed on to the next in pipeline fashion. A delay of three frames, of one-tenth of a second, between an action and a response is acceptable, as long as a new response is generated every thirtieth of a second.

Schedules

Even at the response level, not all processing is immediate. Both input and output events may have duration that requires scheduling. For instance, the system perceives a gesture not as a single instantaneous event, but as a sequence of actions taking place within a fixed span of time. On the output side, a sound response is seldom a single event. At the very least, it has an onset and a termination. Commonly, a single note will require a sequence of amplitude and modulation events. Equally likely is a single response composed of a sequence of discrete sounds. Visually the actions of an animated creature have duration and must therefore be controlled by a schedule.

Global events are scheduled to the nearest second. At the lowest level, the system executes in terms of a response cycle. It inputs and responds. Scheduled inputs are accepted, responses determined and scheduled. Responses that were previously scheduled for this time are output. In simple interactions, the system can accept an input and respond to it in the same clock cycle. In more elaborate interactions the system analyzes the input during one step, decides on the response during the second, and generates the response during the third. Thus, the response occurs three steps after the actual input.

To give a feeling for the schedule-driven events, a few useful schedule entries will be described. Each entry contains the time at which it is to occur, the type of entry, and a pointer to the routine to be executed (Fig. VI.1). The schedule itself is a list of such entries ordered by time of occurrence. Three types of scheduled entry are shown below as examples:

1. A *one-shot* event is executed once and then unscheduled not to be executed again. An example of this is the routine that flashed the strobe light in PSYCHIC SPACE.

2. A *cyclic routine* is scheduled to be executed once every so many clock cycles. Once executed, it is automatically rescheduled to occur again after that number of clock cycles have passed. The cyclic routine that read the floor was scheduled to be called every tenth of a second.

3. The *sequential entry* causes a series of routines to be executed. The entry contains a pointer to a list of routines and the relative time increments at which they are to occur (Fig. VI.2). When the scheduled time arrives, the first routine is executed, the pointer in the entry incremented so it points to the address of the second routine, and the time increment with that address added to the current clock value to become

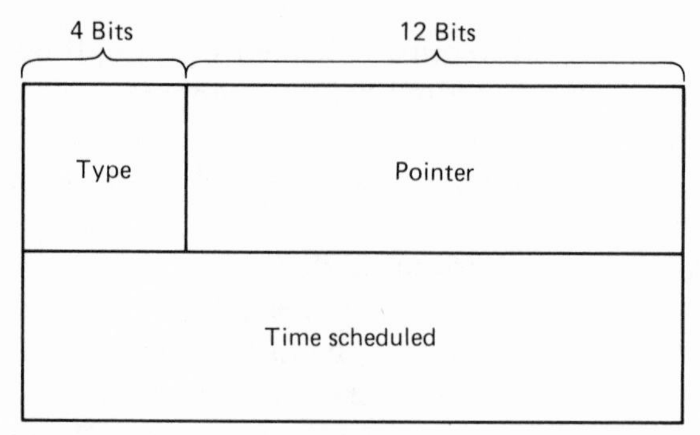

Type: Type of schedule entry
Pointer: Contains address of scheduled routine
Time scheduled: Clocktime at which this entry must be serviced

Fig. VI.1 Schedule entry format

the next time the sequential entry is scheduled to be called. In PSYCHIC SPACE, the series of notes preceding the light activating the phosphors were the result of a single sequential entry in the schedule.

Real-Time Module Structure

When timing imposes a hard constraint, good programming practices may have to be compromised. It is generally good practice to delegate the details of a program to lower- and lower-level procedures (Fig. VI.3). This technique keeps the high level of the program conceptually clean, terse, and easy to understand. However, the process of passing the same parameters to lower and lower levels means that a high overhead penalty is incurred. Parameters must be pushed onto a stack or accessed through indirection. Space must be allocated for the local storage of subordinate routines and registers must be saved.

For instance, "For I = 1,1000 DO Y = Y + F (g(h(I)))" looks compact but includes three hundred procedure calls. It would be faster to apply each function to all the data before passing the results to the next (Fig. VI.4). Organizing

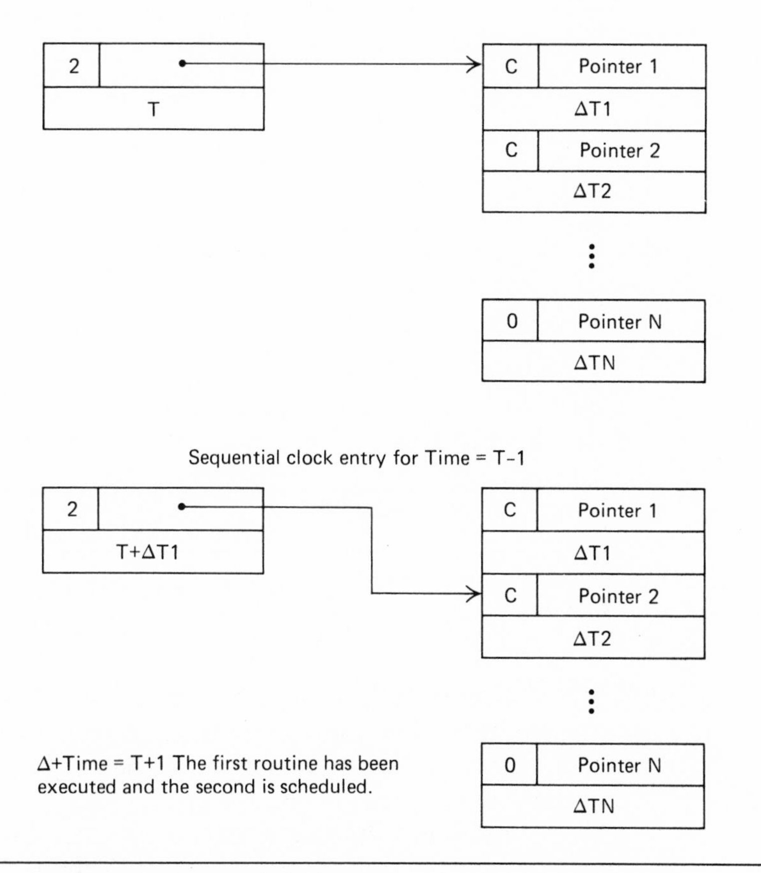

Fig. VI.2 Sequential clock entry

the processing in this way has benefits other than speed. In the original version all of the routines had to be linked together and run on a single processor. In the alternative none of the routines need know anything about the others. They may run on separate processors or be implemented in hardware.

Such routines would be implemented as modules according to a very restrictive set of conventions. In the real-time parts of this system a module is pure code. It is invoked only by the operating software. It does not call other modules and it does not make requests of the system. It executes and simply reports that it is done. Modules may be concatenated

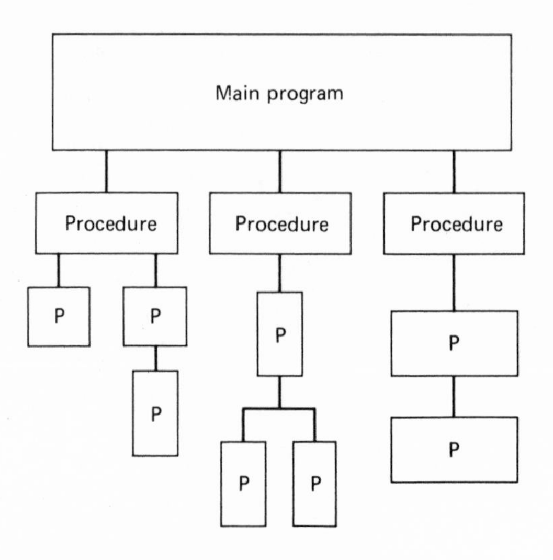

Fig. VI.3 Structured programming modules

to provide composite functions. Where the output conventions of one module are not consistent with the input conventions of the module to be applied to its output, special translation modules will be used.

I have used this technique on a four-processor crosspoint system I designed and implemented while at the University

Fig. VI.4 Real-time module structure

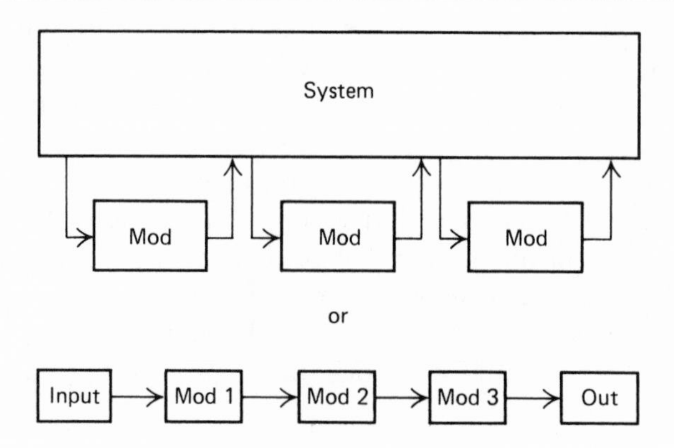

of Wisconsin. Programs were resident on machines. Data and parameters were staged in a crosspoint memory that would suddenly appear in a processor's address space.[7] A single byte sent over a serial line would cause the processor to find the parameter block and execute the appropriate module. The kind of generality that takes time was shunned at all costs. If the system had to check all of its general cases to see which pertained before it did anything, real-time response would be out of the question.

Rather than writing a program to control the system, we want to set up an assembly line of modules, each passing its output on to the next. The modules do not know who is providing this input or getting their output. They just execute.

The process of designing this assembly line is different from programming. The composer considers each context, identifies the modules that will be needed to implement it, the parameters and control data they must be fed, and the processors on which they must run. The composer then creates a control structure that will assure that each module can find its input and knows where to leave its output. Next, the adjacent contexts and the boundary conditions that will trigger transitions to each of them are identified. The boundary conditions are checked by modules dedicated to that purpose. Finally, the modules associated with the context and a table saying what goes where are combined in a single file that is stored on disk.

Currently, these modules and their assembly are pretty much handcrafted, but the longer-range goal is to develop the idea of knowledge-controlled programming. The composer would describe in English the details of the relationship to be defined within a particular context. The system would understand this description in a deep sense although it might request clarification of certain points in English. The system then would identify the modules needed and tie them together. It might even generate graphic sequences and select sound styles from a library automatically. The system would recognize if no existing module, hardware or software, could provide a desired effect and tell the user exactly how to follow the convention required to implement it.

If this dream sounds utopian, it is not entirely. Concrete steps have been taken toward the goal of knowledge-based programming.[8]

Hardware Summary

The general configuration of the system is designed to allow the evolution of a hierarchical structure of disparate processors with an emphasis on high speed transfer. It is to be a closely coupled system, where command, control, and allocation for one level are always accomplished at a higher level. There will be a main interface bus under the control of the staging processor. This bus will have a very large address space and not be used for computations. Separate buses will be created where bandwidth requirements are great. For instance, video control signals, including line number, pixel number, and pixel values, are all bussed for both pattern recognition and graphic interfaces. Thus a number of feature detectors and graphics generators can operate in parallel on the same input data. Subordinate processors are connected by means of dual-port memories and special intervention hardware that assures that the processor bus is quiescent before a memory swap is attempted and alerts the processor that new instructions await (Fig. VI.5). The development philosophy is that algorithms will first be developed in software on central processors and once perfected, delegated to subordinate microprocessors and finally to microsequencers and hardwired logic. Additional high-bandwidth data paths will be connected as needed from one part of the system to another. However, although this communication is ostensibly from one subprocessor to another, the significance of the transfer is provided to both sender and receiver from above.

SAP — The Small Address Problem

A dominant concern of any system designer working with small processors is the limited address space accessible by sixteen-bit machines. The problem arises because the sixteen-

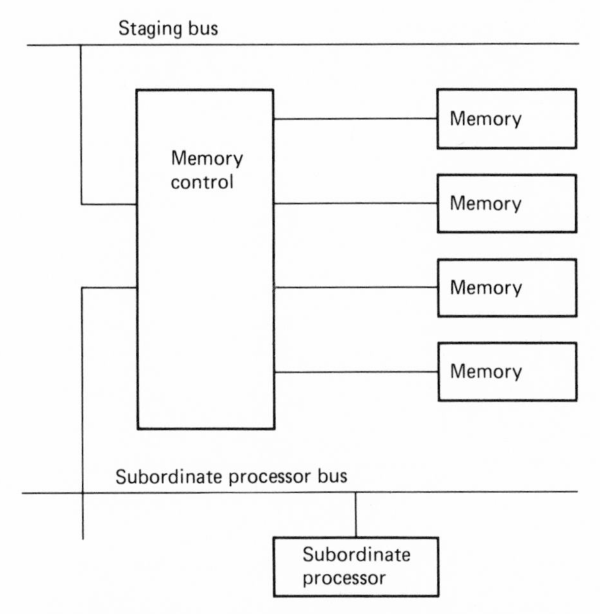

Fig. VI.5 Staging subordinate processor through dual-port memory

bit word allows access to only two to the sixteenth power (2^{16}) = 65,536 bytes of memory. Therefore, even though memory costs less and less, the smaller systems are unable to take advantage of it. Since any significant program exceeds this limit, the constraint poses a real hardship.

The classic solution is memory mapping; special hardware expands the processor's address space to eighteen, twenty, or twenty-four bits. This is done by concatenating high-order address bits stored in mapping registers with the low-order bits in the computer's address (Fig. VI.6). At any moment the mapping registers can access only 64,000 bytes. However, by changing the contents of the mapping registers, the processor's address space can point at different areas of

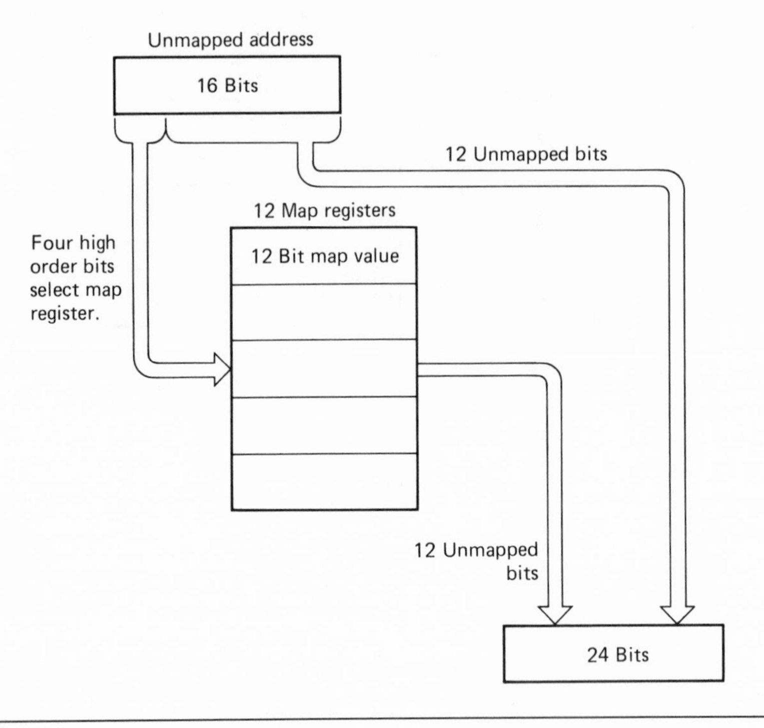

Fig. VI.6 Mapped address generation

the system space at different times (Fig. VI.7). Unfortunately, memory mapping creates as many problems as it solves. There is considerable overhead associated with its use and an amazing quantity of computer time and programmer effort can be consumed fighting with the mapping hardware.

In the low-level processors and feature detectors, useful work can be accomplished in a limited memory. In fact, since each processor has its own address space, parallel processing serves to increase the system address space as well as its throughput. However, at higher levels in the system, the address-space limitation is intolerable and so memory mapping or some other alternative must be used. In particular, it is reasonable to have the data and code required of all adjacent contexts staged in semiconductor, as opposed to disk memory so that when a context switch occurs, it is a matter of memory

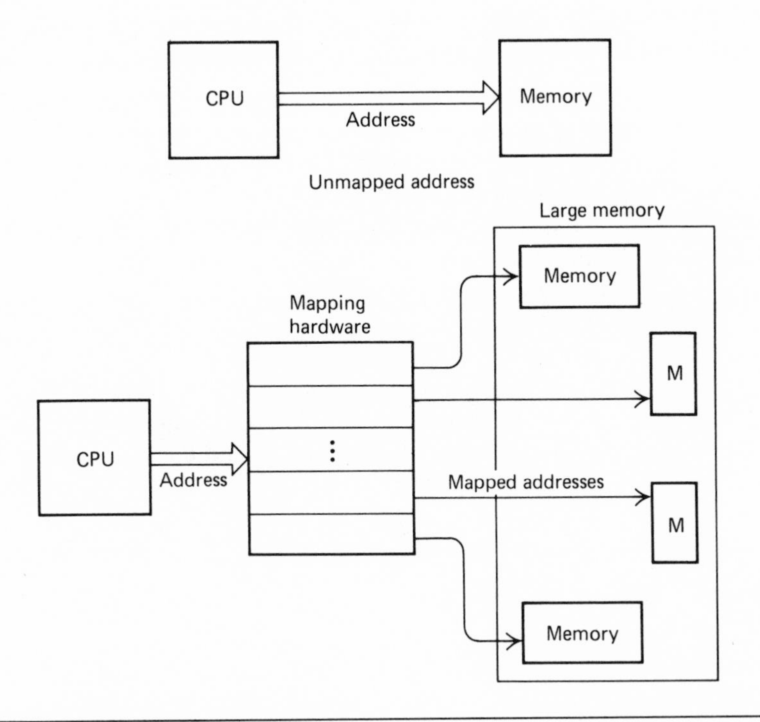

Fig. VI.7 Memory mapping

speed transfers rather than disk access speed. For this reason, in the current incarnation of the system, the sixteen-bit staging processor controls transfers on the twenty-four address-bit staging bus through memory mapping. This memory mapping scheme provides sixty-four separate contexts, each with fifteen mapping pages and sixteen much smaller device pages.

NOTES

1. R. M. Metclafe & D. R. Boggs, "Ethernet: Distributed Packed Switching for Local Computer Networks," *Communications of the Association for Computing Machinery* 19, No. 17 (July 1976):396–403.

2. S. M. Ornstein, W. R. Corwther, M. F. Kraley, R. D. Bressler, A. Michel & F. E. Heart, "Pluribus — a Reliable Multiprocessor" (AFIPS), 44 (1975).

3. Mark Dowson, "The Demos Multiple Processor Technical Summary," *National Physical Laboratory Report* 102 (April 1978).

4. Larry Wittie, "MICRONET: A Reconfigurable Network for Distributed Systems Research," *Simulation* (September 1978):145–153.

5. R. J. Swann, S. H. Fuller, & D. P. Siewiorek, "CM* — A Modular Multiprocessor" (AFIPS), 44 (1975).

6. S. Fuller, R. Swan & W. Wulf, "The Instrumentation of C. MMP: a Multi-Mini Processor," *IEEE COMPCON* (1973).

7. M. W. Krueger et al, "Innovative Video Applications in Meteorology (IVAM)," Technical Report, Space Science and Engineering Center (Madison: University of Wisconsin, 1978).

8. M. W. Krueger, R. E. Cullingford, & D. A. Bellavance, "Control Issues in a Multiprocess Computer Aided Design System Containing Expert Knowledge," *Proceedings International Conference on Cybernetics and Society* (October 1981).

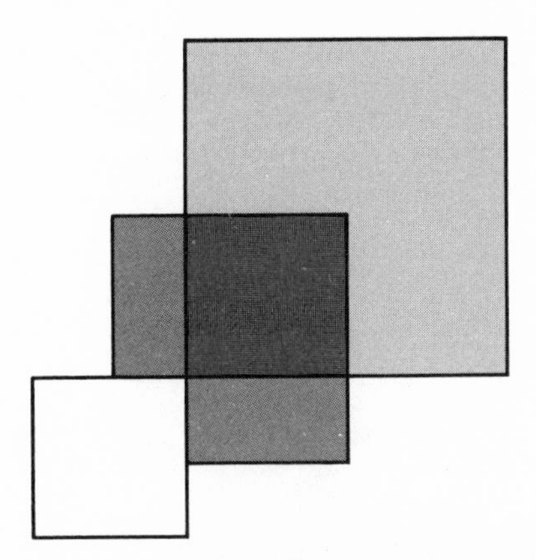

APPENDIX VII
GLOWFLOW SOUND

The model planned was based on statistical variation around a compositional unit called a major cycle. The major cycle was divided into a number of smaller units called minor cycles about one second in duration and weighted according to the likelihood of sound during the period. Similarly, each minor cycle was further divided into a given number of smaller time increments, each of which was again assigned a weight for the probability of a change in the analog control voltage from the computer to the Moog (Fig. VII.1). If there were to be attacks during a given minor cycle, the timbre routine went through a network of weighted decisions that chose the waveform and type of modulation (Fig. VII.2). One of three decay rates was also selected. These decays were the only envelopes available. Choices of analog voltages were based on a probability distribution associated with the previous value. Due to the memory limitations of the PDP-12 only eight discrete voltage levels were used (Fig. VII.3). The events of a given minor cycle were repeated from one to five

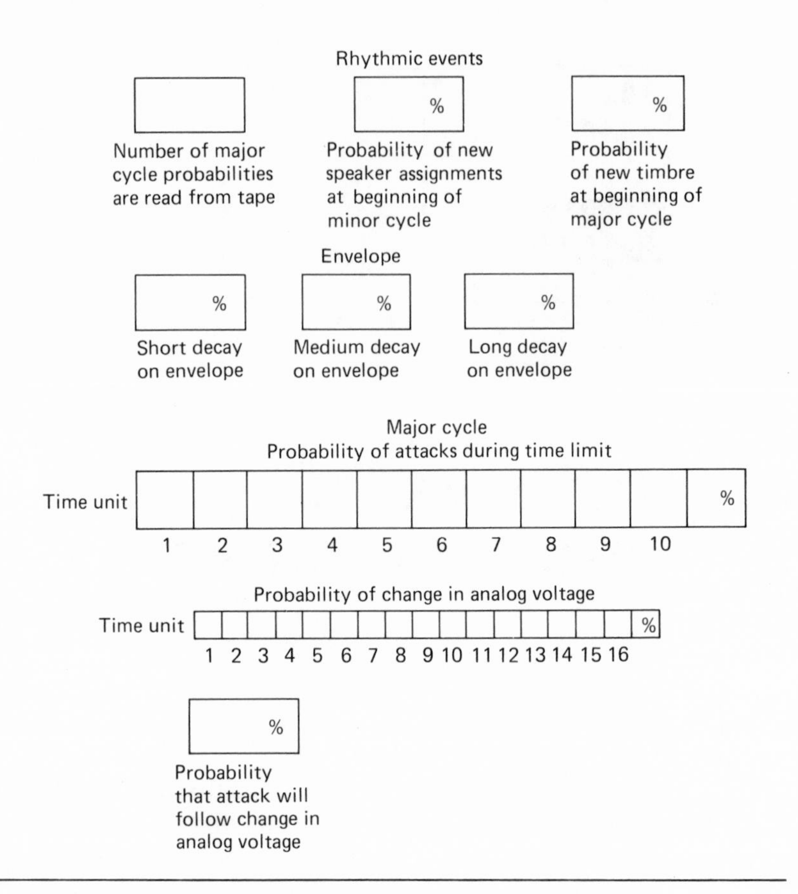

Fig. VII.1 Rhythmic events

times depending on another distribution called the redundancy vector. (The score for one of the compositions is included as figures 1, 2, and 3.)

The responsiveness added to this program either varied the pitch range being used or forced an attack when a person was on one of the mats. It was also possible to make other subtle changes such as adding or deleting white noise or reverb.

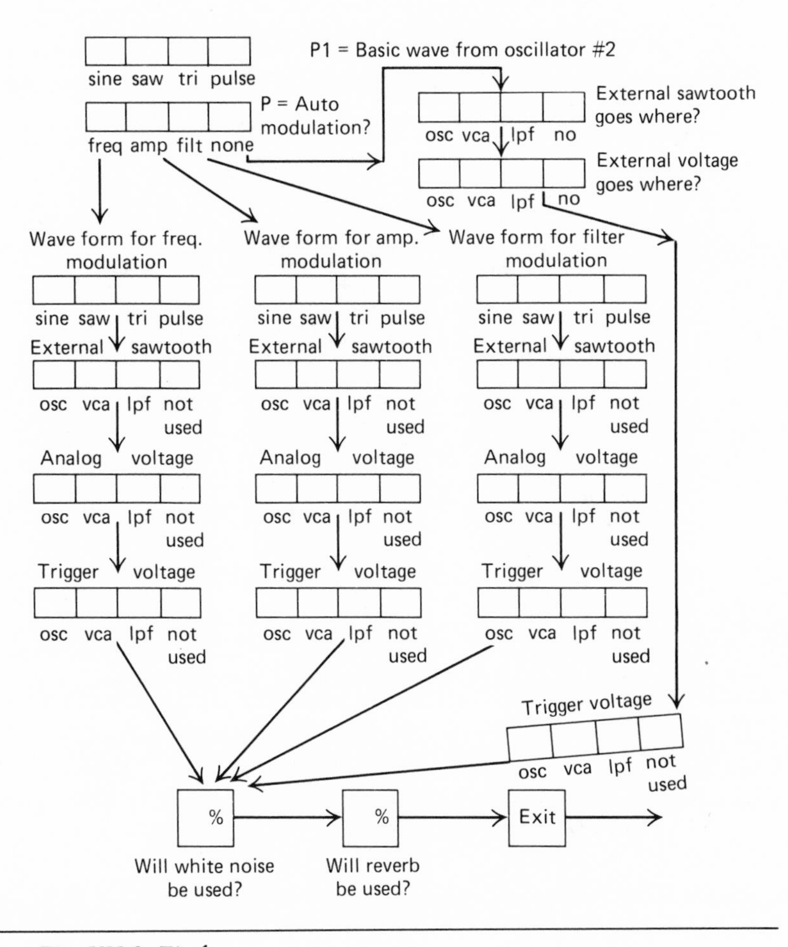

Fig. VII.2 Timbre

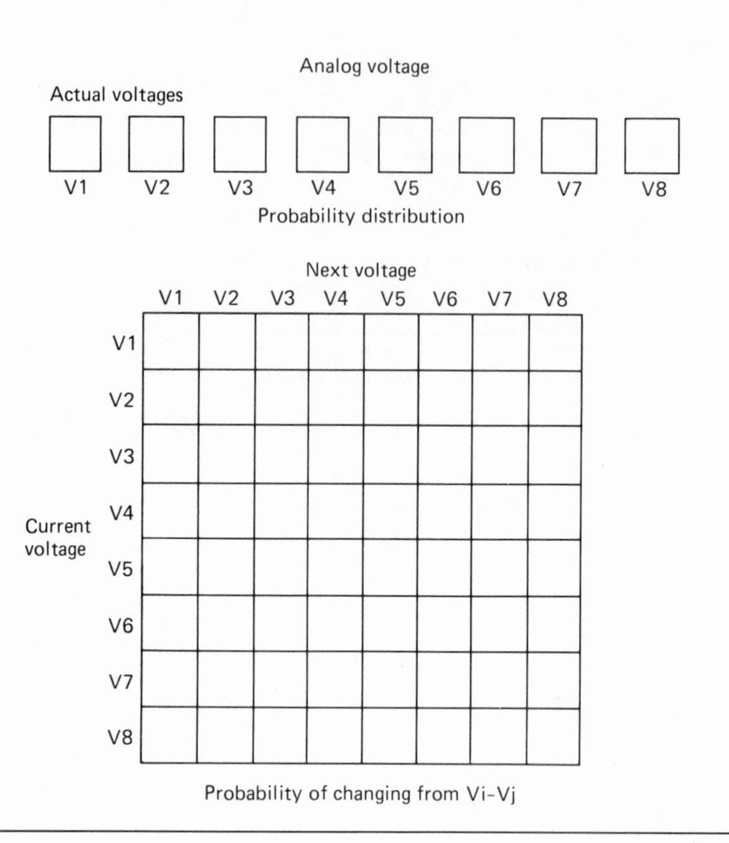

Fig. VII.3 Analog voltage

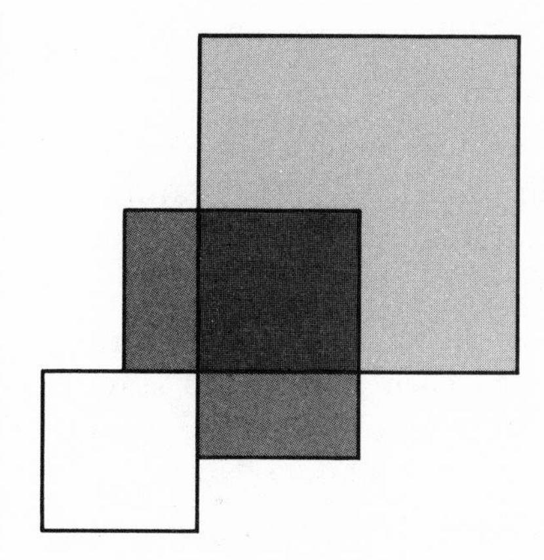

GLOSSARY

Accumulator The central register in older machines. The only one where additions and logical operations could be applied.

Adage Vector graphic display system manufactured by Adage, Inc. of Boston, Massachusetts.

AI Abbreviation for Artificial Intelligence.

Alphanumeric Refers to the magnitude of an electronic signal. The greater the amplitude, the higher the voltage, the louder the sound, or the brighter the image.

Analog As opposed to digital. See Appendix I. The branch of computer science concerned with duplicating.

Artificial Intelligence Those human skills that illustrate our intelligence, e.g., understanding visual images, understanding speech and written text, problem solving, medical diagnosis.

Analog-to-digital converter Converts an analog signal representing some real-world quantity into a binary number that approximates its current value.

Architecture This term is used to refer to the high-level organization of a hardware or software system.

Associative memory A sophisticated form of memory that combines logic circuitry with each item of memory. This feature departs from the tradition of separating processing from memory and can greatly accelerate memory searches.

Batch processing The practice of running a stream of programs through the computer such that each is executed start to finish without interruption. As opposed to multiprogramming or time sharing.

Bit A unit of binary information containing either a zero or a one.

Bus A means of distributing a set of signals so the computer can be interfaced with memory and external devices. Any device that is physically compatible with the bus and observes the protocols that it demands should work immediately after being plugged in.

Byte Eight bits of binary information. Can be used to represent a single character, e.g., 01000001 = 'A'.

Capacitance A measure of the ability to store electrical charge or resist a change in voltage.

Clock A clock is a square wave that is used as a source of synchronization in a digital circuit.

Clocked A circuit or signal or event which is controlled by a clock is said to be clocked.

Code The text of a computer program readable either by people or by machine.

Colorizing A method for arbitrarily assigning colors to image data. For instance, a different color can be assigned to each grey level in a black-and-white image.

Concatenate To place end to end or link together. The result of concatenating 'dfdfdf' with 'abab' is 'dfdfdfabab'.

Conceptual art A recent movement in art where a work is to be appreciated for its conceptual qualities as well as, or in lieu of, its perceptual ones.

Counter A digital device whose outputs count in binary each time a clock pulse occurs on their clock input.

CPU Central Processing Unit. The part of the computer that controls the fetching of instructions from memory, the interpretation of those instructions, and their execution.

CRT Cathode Ray Tube. The tube in a television set.

DAC Digital-to-Analog Converter.

Dada An art movement in the early part of this century that poked fun at the traditional view of art.

Data tablet A device for putting two-dimensional information into the computer. The position of a stylus held by the user is digitized and continually fed to the computer as the user moves it about a special surface.

Differentiation A mathematical procedure for finding the equation of a curve whose value at each point is equal to the slope of a given function.

Digital See Appendix I on Digital Fundamentals.

Digital-to-analog converter Device for translating a digital quantity, represented as a number of binary bits, into a voltage or current whose magnitude is in some way equivalent.

Digitize To convert an analog quantity into its binary equivalent.

Doppler shift The shift in frequency perceived when a frequency source is moving with respect to a fixed reference. The classic example is a train whistle, which appears to increase in pitch as the train approaches and decrease in pitch as it moves away.

Dual-port A dual-port device can be physically connected to two computer busses simultaneously. Access to the device by one bus does not require the loss of bus cycles on the other.

EEG Electroencephalogram. A recording of the electrical signals generated by brain activity.

Electroluminescence Similar to phosphorescence or fluorescence except that the emission of light is stimulated by an electrical signal.

Electrostatic Related to electrical phenomena that result from the storage of a static charge as opposed to the flow of current through a circuit. The force that attracts a hair to a comb is electrostatic. Such forces can be used to bend an electron beam inside a cathode ray tube. While television usually employs electromagnetic forces for that purpose, oscilloscopes often use electrostatic deflection.

EMG Electromyogram. Recording of electrical signals generated by muscle activity.

EKG Electrocardiogram. Recording of electrical signals generated by the functioning of the heart muscle.

Envelope The waveform used to control the amplitude of another waveform.

Factorial The factorial of an integer n is formed by multiplying together all the integers from one to n.

Femptosecond One-quadrillionth of a second.

Flashtube A gas discharge tube used for generating very brief, very bright flashes of light, e.g., a strobelight.

Frequency The number of cycles per unit of time. Corresponds to the notion of pitch in music.

Fundamental The lowest tone in a complex sound. The others are called harmonics.

GSR Galvanic Skin Resistance. Skin resistance changes in response to stress.

Harmonics Components of a complex sound other than the fundamental. The frequencies of the harmonics are integral multiples of the fundamental.

Hidden line removal In computer graphics, the process of eliminating lines that would be invisible to the viewer because they are occluded by surfaces that are closer to the viewer.

Hologram A film image created and viewed with the help of a laser beam. The hologram records a window in a three-dimensional scene. By moving one's head, one can change the point of view of the three-dimensional scene.

Impressionists A group of painters of the late nineteenth century who sought to capture the dynamic act of perception as opposed to the static photographic recording of a scene.

Infrared The part of the electromagnetic spectrum characterized by waves of lower frequency and longer wavelengths than those of visible light.

Infrasound Sounds whose frequencies lie below the range of human hearing.

Integration The process for finding a function whose derivative is the given function.

Interface The process of making an interconnection between two pieces of hardware or software. A device or piece of software that accomplishes such an interconnection.

Interlace In television, the practice of dividing an image into two fields, one with odd-numbered lines, the other with even-numbered lines. The two fields are transmitted and displayed sequentially.

Interpolation The process of using known values to calculate unknown values that lie between them.

Intersection In set theory, points that are contained in two sets comprise their intersection.

Josephson Junction A very advanced form of switching technology that may be the basis for the digit circuits of the future. Switching times are as low as ten picoseconds (trillionths of a second), as much as a hundred times faster than current devices.

Klein bottle A bottle in which the inside and outside surfaces are the same. Analogous to a moebius strip.

Logic Digital circuits are designed in terms of primitive functions that are identical to those used in the formal logic of philosophers and mathematicians. Hence, collections of such circuitry are often referred to as logic.

Loop A programming construction that allows a programmer to tell the computer to execute a particular sequence of instructions a specified number of times or until a particular condition is met.

Mass storage Usually refers to high-capacity tape or disk storage.

Megabyte A million bytes.

Megahertz A million cycles per second.

Microprocessor A computer implemented as a tiny integrated circuit smaller than a fingernail. The small size, low-power requirement and low cost of microprocessors have revolutionized computer applications.

Microsecond A millionth of a second.

Microsequencer A primitive control unit that is not as general as a computer. Today, most computers are implemented with microsequencers.

Millisecond One thousandth of a second.

Modulation The use of one signal to control some aspect of another, such as the frequency or amplitude.

Monitor A video display that accepts video as opposed to RF input, i.e., a monitor that has no tuner and cannot receive broadcast signals. It can only receive signals locally generated from a camera or video tape recorder.

Moog The first commercially successful electronic sound synthesizers were designed by Robert Moog.

Multiprogramming The practice of running several programs simultaneously on the same computer. The advantage is that while one program may be waiting for a slow input/output process, another can be executing.

Nanosecond One billionth of a second.

Operating system A master control program that allows other programs to run and provides them with utility services as they are running.

Oscilloscope An instrument for displaying electrical waveforms.

Partial animation A crude form of animation that employs as few as two or three frames per second as opposed to the twenty-four frames per second required for full animation.

Passive art form In this book "passive" refers to the fact that the audience does not contribute to the realization of the work. Even though dance involves activity on the part of the performers, people in the audience are passive consumers of the performance.

Passive sensor A sensor that simply accepts incoming signals is passive. An ultrasonic system is active because it actively transmits a signal that is reflected off the participant's body and then sensed.

Pattern recognition The branch of computer science that deals with the interpretation of visual images and other patterns.

PDP-11 A sixteen-bit minicomputer manufactured by Digital Equipment Corporation.

PDP-12 An earlier twelve-bit minicomputer manufactured by Digital Equipment Corporation.

Phase A particular point in a periodic waveform.

Phosphorescent A material that absorbs light when exposed to it and then continues to emit light after the exciting source is removed, is said to be phosphorescent.

Photocell An electrical device for detecting light.

Picosecond One trillionth of a second.

Pipelining The practice of creating an assembly line of processing elements. The output of one processor becomes the input of the next and so on. All processors are continually in operation.

Pixel A point in a raster graphics image.

PLATO A computer-aided instruction system developed at the University of Illinois.

Plotter A device for making drawings on paper.

Polyethylene A thin plastic film that comes in large sheets.

Problem solver A computer program that is able to reason from the statement of a problem and the means available to solve it, the sequence of steps required to create a solution.

PROM A programmable read-only memory. This device can be used to store binary information in a form accessible to digital circuitry. The information is stored permanently and not affected when power is removed. It is programmable only in the sense that it is blank when purchased and must be programmed by the end user. Once programmed, it can never be changed.

Raster The organization of a television image into scan lines is called a raster.

RGB Red, green, blue. A method of dealing with a color television image as three separate signals rather than combining them into a single signal as NTSC encoding does.

Real-time Is applied to a system that computes its results as fast as needed by a real-world system. In this book, the term means that there is no perceptible delay to the human observer. In general use the term is often perverted to mean within the patience and tolerance of a human user.

Rear-project To project an image on one side of a translucent screen and to have the viewer stand on the other side. Thus, the viewers are unaware of the projector and cannot block its beam with their bodies.

Register A collection of one-bit storage devices that are used for temporary storage inside a computer or some other digital circuit.

Sawtooth An electronic waveform that looks like a sawtooth. It is characterized by a linear increase in voltage that falls to zero instantaneously when the peak value is attained.

Scale a graphic object Make it smaller by multiplying the coordinates of each to the points that define the object by some constant value.

Scale a waveform To reduce its amplitude by a constant factor.

Semantic Having to do with meaning and significance as opposed to syntactic which has to do with pure structure.

Sensory deprivation The total denial of sensation.

Sine See Appendix III on electronic sound.

Skinner Box A device invented by B. F. Skinner for shaping animal behavior.

Soft copy Information stored in the computer that can be used to generate a display on the computer screen. This information can also be used to print a hard copy on paper.

Solenoid An electromechanical device for translating an electrical signal into an on/off mechanical action.

Solid state Employing transistors (the product of solid state physics) instead of vacuum tubes.

Spectrum analysis The breakdown of a complex waveform into its constituent frequencies.

Square wave A periodic binary waveform that is always either zero or one.

Staircasing The visual effect created when a step function is used to approximate a continuous waveform.

Strain gauge A sensor that detects deformations of a piezoelectric crystal.

Synaesthesia The combination of stimuli and sensation so that the normal boundaries of perception are broken down.

Syntactic Having to do with the grammar or structure of information as opposed to its meaning or significance, as opposed to semantic.

Time-sharing The practice of allowing a number of people to use a computer simultaneously. For certain functions, such as word processing, the computer is far faster than the human user. Thus, it can handle the needs of one user while the others are moving their fingers to type the next character or scratching their heads to decide what to do next.

Topologists Mathematicians concerned with the most general properties of surfaces.

Track ball A graphic input device that the user controls by rolling a stationary ball mounted in a box. To move a cursor on the screen, the user rolls it in the direction he wants it to move.

Transducer A device that translates one kind of energy into another. A speaker, a doorbell, a light bulb, and a hairdryer are all examples of transducers.

Triangulation A method of pinpointing position by computing direction with respect to three known locations.

Triangle wave An electronic waveform that looks like a triangle. It increases linearly to a peak and then decreases at the same rate.

Ultrasonic Refers to sounds that are at higher frequencies than the human ear can hear.

Ultrasound Ultrasonic sound.

Waveform The shape a signal has when its amplitude is plotted with respect to time.

Zoom To make the viewer appear to move closer to a scene by changing the focal length of the lens. To accomplish the same visual result by means of computation in computer graphics.

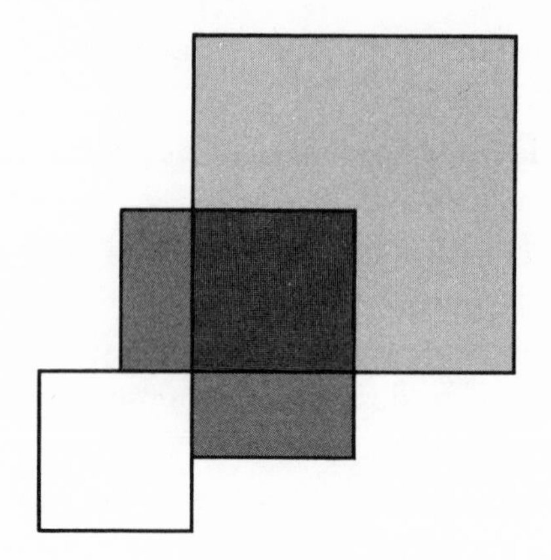

BIBLIOGRAPHY

Alles, H. G. "Music Synthesis Using Real Time Digital Techniques." Tech. Report, Bell Laboratories, Murray Hill, N.J.

Badler, Norman, et al. "Digital Representations of Human Movement." *Computing Surveys* 11, no. 1, March 1979.

Baecker, Ronald, et al. "Towards a Laboratory for Motion Analysis." *ACM SIGGRAPH Proceedings,* 1981.

Batter, James J., et al. "Grope-1." *IFIPS Proceedings* 71, 1972.

Bitzer, D. L., et al. "PLATO: A Computer Based System Used in the Engineering of Education." *IEEE Proceedings* 59, June 1971.

Bolt, Richard. "Gaze Orchestrated Dynamic Windows." *ACM SIGGRAPH Proceedings,* 1981.

———. "Put That There: Voice and Gesture at the Graphic Interface." *ACM SIGGRAPH Proceedings,* 1980.

Booth, Taylor L. *Digital Networks and Computer Systems.* New York: Wiley, 1978.

"A Building That Moves in the Night." *New Scientist,* 19 March 1981.

Burtnyk, N., et al. "Computer Animation of Free Form Images." *ACM SIGGRAPH Proceedings* 9, no. 1, 1975.

Butler, Samuel. *Erewhon.* New York: Penguin, 1970.

Brooks, Frederick P. "The Computer 'Scientist' as Tool-smith — Studies in Interactive Graphics." *IFIP* Congress Proceedings, North-Holland Publishing Co., 1977.

Cage, John. *Silence.* Cambridge, Mass.: MIT Press, 1967.

"Chess Computers Start to Give Humans a Tough Game." *Electronic Engineering Times,* 18 April 1977.

"Chess 4.7 Versus David Levy: The Computer Beats a Chess Master." *Byte,* December 1978.

Clark, James. "A VLSI Geometry Processor for Graphics." *Computer,* July 1980.

"C-Mos Implant to Aid Deaf." *Electronics,* 20 February 1975.

Cohen, Harold. "What Is an Image." *IJCAI,* 1979.

Colby, Kenneth M., et al. "Artificial Paranoia." *Artificial Intelligence* 2, no. 1, 1971.

Csuri, C. "Toward an Interactive High Visual Complexity Animation System." *ACM SIGGRAPH Proceedings,* 1979.

Cullingford, Richard, et al. "Automated Explanations as a Component of a CAD System." To appear in *IEEE Transactions on Systems, Man and Cybernetics,* special issue on human factors and user assistance in CAD, 1982.

Davis, Randal, et al. "Production Rules as a Representation for a Knowledge-Based Consultation Program." *Artificial Intelligence* 8, 1977.

Devey, G. B., et al. "Ultrasound in Medical Diagnosis." *Scientific American* 238, May 1978.

Dewey, John. *Experience and Education.* New York: Macmillan, 1963.

Donelson, William. "Spatial Management of Information." *ACM SIGGRAPH Proceedings* 12, no. 3, 1978.

Dowson, Mark. "The Demos Multiple Processor Technical Summary." *NPL Report Com.* 102, April 1978.

Dreyfus, H. L. *What Computers Can't Do.* New York: Harper & Row, 1972.

———. "Why Computers Must Have Bodies in Order to Be Intelligent." *Review of Metaphysics* 21, 1967.

Duff, M. J. B. "CLIP4: A Large Scale Integrated Circuit Array Parallel Processor." *International Joint Conference on Pattern Recognition Proceedings,* 1976.

Ehora, Theodore. "World Chess Championship Computer." *Creative Computing,* January 1979.

"An Electronic Link to the Visual Cortex May Let Blind 'See'." *Electronics,* 20 December 1973.

"Focus on Light." Trenton, N.J.: N.J. State Museum Cultural Center, 20 May–10 September 1967.

Foldes, Peter. "Hunger (Le Faim)." National Film Board of Canada, Learning Corporation of America, 1974.

Fuller, S., et al. "The Instrumentation of C.mmp: A Multi-Mini Processor." *IEEE COMPCON,* 1973.

Geldzahler, Henry. "Happenings: Theater by Painters." *Hudson Review* 18, no. 4, Winter 1965–66.

Hayes-Roth, F., et al. "Focus of Attention in the HEARSAY II Speech Understanding System." *International Joint Conference on Artificial Intelligence,* 1977.

Jackson, J. H. "Dynamic Scan-Converted Images with a Frame Buffer Display Device." *ACM SIGGRAPH Proceedings,* 1980.

"John Whitney Interview Conducted by R. Brick." *Film Culture* 53–7:39–83, March 1979.

Johnson, Avery. "The Impact of Computer Graphics on Architecture." In *Computer Graphics in Architecture and Design,* edited by Murray Milne. New Haven: Yale University Press, 1968.

Kingsley, E. C., et al. "SAMMIE — A Computer Aid for Man-Machine Modelling." *ACM SIGGRAPH Proceedings,* 1981.

Krueger, M. W., et al. "Control Issues in a Multiprocess Computer Aided Design System Containing Expert Knowledge." *Proceedings International Conference on Cybernetics and Society,* October 1981.

Krueger, M. W., et al. "Innovative Video Applications in Meteorology (IVAM)." Technical Report, Space Science and Engineering Center, University of Wisconsin, Madison, 1978.

Lawrence, L. George. "Communications Via Touch." *Electronics World,* May 1968.

Lippard, Lucy. *Six Years: The Dematerialization of the Art Object from 1966–1972.* New York: Praeger, 1973.

Lipscomb, J. S. "Three-Dimensional Cues for a Molecular Computer Graphics System." Ph.D. dissertation, University of North Carolina at Chapel Hill, 1979.

Ludwig, Arnold. "'Psychedelic' Effects Produced by Sensory Overload." *American Journal of Psychiatry* 128, April 1972.

McLuhan, H. Marshall. *The Medium Is the Massage.* New York: Random House, 1967.

Marcus, Aaron. "Experimental Visible Languages." Proceedings of *Apollo Agonistes: The Humanities in a Computerized World,* Vol. 2, SUNY, Albany, 1979.

Marsh, Peter. "The Making of the Computerized Car." *New Scientist,* 6 December 1979.

Marshall, Robert, et al. "Procedure Models for Generating Three Dimensional Terrain." *ACM SIGGRAPH Proceedings,* 1980.

Meehan, James R. "TALE-SPIN, An Interactive Program That Writes Stories." *Fifth International Joint Conference on Artificial Intelligence,* vol. 1, 1977.

Metclafe, R. M., et al. "Ethernet: Distributed Packed Switching for Local Computer Networks." *CACM* 19, no. 7, July 1976.

"A Method of Interactive Visualization of CAD Surface Models on a Color Video Display." *ACM SIGGRAPH Proceedings,* 1981.

Miller, Barry. "Remotely Piloted Aircraft Studied." *Aviation Week and Space Technology* 92, no. 22, 1 June 1970.

Moore, O. K. "O.K.'s Children." *Time* 76, 7 November 1960.

Nelson, Max. "Vectorized Procedure Models for Natural Terrain: Waves and Islands in the Sunset." *ACM SIGGRAPH Proceedings,* 1981.

The Neuroprostheses Program. "Data Processing, LSI Will Help to Bring Sight to the Blind." *Electronics,* 24 January 1974.

Newman, W. M., et al. *Principles of Interactive Computer Graphics.* New York: McGraw-Hill, 1979.

Noll, Michael A. "Computer Animation and the Fourth Dimension." *Fall Joint Computer Conference,* AFIPS, pp. 33–42, 1977.

The Notebooks of Leonardo da Vinci. J. P. Richter, ed. New York: Dover, 1970.

O'Donnell, T. J., et al. "GRAMPS — A Graphics Language Interpreter for Real-Time Interactive Three Dimensional Picture Editing and Animation." *ACM SIGGRAPH Proceedings,* 1981.

Ornstein, S. M., et al. "Pluribus — A Reliable Multiprocessor." AFIPS, vol. 44, 1975.

Packard, Edward. *The Cave of Time.* New York: Bantam Books, 1979.

Parke, F. I. "Computer Generated Animation of Faces." University of Utah Computer Science Department, UTEC-CSc-72-120, 1970. NTIS Ad-762 022. Abridged version in *Proceedings ACM National Conference,* 1972.

Payne, Doug. "Bye-bye Buzby, Bye-bye." *New Scientist,* 28 May 1981.

"Planes, Tools, Even Chairs . . . Designed by Computer." *Popular Science,* February 1981.

Raphael, B. *The Thinking Computer.* San Francisco: W. H. Freeman, 1976. *IEEE International Convention Record,* part 9, 1967.

Reeves, William. "In Betweening for Computer Animation Utilizing Moving Point Constraints." *ACM SIGGRAPH Proceedings,* 1981.

Reichardt, Jasia. "Art at Large." *New Scientist,* 4 May 1972.

Robinson, A. L. "Computer Films: Adding an Extra Dimension to Research." *Science* 200, May 1979.

Schank, R., et al. *Scripts, Plans, Goals and Understanding.* Hillsdale, N.J.: Erlbaum Press, 1977.

Shoup, Richard G. "Color Table Animation." *ACM SIGGRAPH Proceedings,* 1979.

Skinner, B. F. *Science and Human Behavior.* New York: Macmillan, 1953.

"Simple Teletext Keypad." *Popular Science,* January 1981.

Smith, Seward, et al. "Stimulation Seeking During Sensory Deprivation." *Perceptual and Motor Skills* 23, no. 3, part 2, 1966.

Software. New York: Jewish Museum, 1970.

Stern, Garland. "Softcel — An Application of Raster Scan Graphics to Conventional Cel Animation." *ACM SIGGRAPH Proceedings,* 1979.

Stewart, Clifford D. "Integration of Interactive Graphics in the Real-time Architectural Process." *Online 72.* Middlesex, England: Paca Press, September 1972.

Sutherland, Don. "Electronic Clones Are Coming: Will the Real Peter Fonda Please Stand Up?" *Popular Photography,* December 1976.

Sutherland, Ivan E. "A Head Mounted Three Dimensional Display." *Fall Joint Computer Conference* AFIPS, vol. 33-1, 1968.

Swann, R. J., et al. "CM* — A Modular Multiprocessor." AFIPS, vol. 44, 1975.

Swindell, W. "Computerized Tomography: Taking Sectional X-rays." *Physics Today* 30, December 1977.

Teja, Edward. "Marrying Voice Recognition and Synthesis, Robot Pontificates on Presidential Race." *EDN,* 5 August 1980.

Tsugawa, S., et al. "Three Dimensional Movement Analysis of Dynamic Line Images." *IJCAI,* 1979.

Walter, P. E. "Computer Graphics Used for Architectural Design and Costing." In *Computer Graphics,* edited by R. D. Parslow et al. London and New York: Plenum Press, 1969.

Weiman, Carl. "Continuous Anti Aliased Rotation and Zoom of Raster Images."*ACM SIGGRAPH Proceedings,* 1980.

Weinberg, R. "Computer Graphics in Support of Space Shuttle Simulation." *ACM SIGGRAPH Proceedings* 12, no. 3, 1978.

Weiner, D. "Test Tubetelevision: at WNET's Experimental Workshop." *American Film* 4:33–34, March 1979.

Weizenbaum, Joseph. "Contextual Understanding by Computers." *Communications of the ACM* 10: August 1967.

Winston, Patrick. *Artificial Intelligence.* Reading, Mass.: Addison-Wesley, 1977.

———., ed. *The Psychology of Computer Vision.* New York: McGraw-Hill, 1975.

Wittie, Larry. "MICRONET: A Reconfigurable Network for Distributed Systems Research." *Simulation,* September 1978.

INDEX